EDUCATION IN A COMPETITIVE AND GLOBALIZING WORLD SERIES

COMPUTER-ASSISTED TEACHING: NEW DEVELOPMENTS

EDUCATION IN A COMPETITIVE AND GLOBALIZING WORLD SERIES

Success in Mathematics Education
Caroline B. Baumann
2009. ISBN: 978-1-60692-299-6

Mentoring: Program Development, Relationships and Outcomes
Michael I. Keel (Editor)
2009. ISBN: 978-1-60692-287-3

Mentoring: Program Development, Relationships and Outcomes
Michael I. Keel (Editor)
2009. ISBN: 978-1-60876-727-4 (Online Book)

Motivation in Education
Desmond H. Elsworth (Editor)
2009. ISBN: 978-1-60692-234-7

Evaluating Online Learning: Challenges and Strategies for Success
Arthur T. Weston (Editor)
2009. ISBN: 978-1-60741-107-9

Enhancing Prospects of Longer-Term Sustainability of Cross-Cultural INSET Initiatives in China
Chunmei Yan
2009. ISBN: 978-1-60741-615-9

Reading at Risk: A Survey of Literary Reading in America
Rainer D. Ivanov
2009. ISBN: 978-1-60692-582-9

Reading: Assessment, Comprehension and Teaching
Nancy H. Salas and Donna D. Peyton (Editors)
2009. ISBN: 978-1-60692-615-4

Reading: Assessment, Comprehension and Teaching
Nancy H. Salas and Donna D. Peyton (Editors)
2009. ISBN: 978-1-60876-543-0 (Online Book)

Multimedia in Education and Special Education
Onan Demir and Cari Celik
2009. ISBN: 978-1-60741-073-7

Rural Education in the 21st Century
Christine M.E. Frisiras (Editor)
2009. ISBN: 978-1-60692-966-7

Nutrition Education and Change
Beatra F. Realine (Editor)
2009. ISBN: 978-1-60692-983-4

The Reading Literacy of U.S. Fourth-Grade Students in an International Context
Justin Baer, Stéphane Baldi, Kaylin Ayotte,Patricia J. Gree and Daniel McGrath
2009. ISBN: 978-1-60741-138-3

Teacher Qualifications and Kindergartners' Achievements
Cassandra M. Guarino, Laura S. Hamilton, J.R. Lockwood, Amy H. Rathbun and Elvira Germino Hausken
2009. ISBN: 978-1-60741-180-2

PCK and Teaching Innovations
Syh-Jong Jang
2009. ISBN: 978-1-60741-147-5

IT- Based Project Change Management System
Faisal Manzoor Arain and Low Sui Pheng
2009. ISBN: 978-1-60741-148-2

Learning in the Network Society and the Digitized School
Rune Krumsvik (Editor)
2009. ISBN: 978-1-60741-172-7

Effects of Family Literacy Interventions on Children's Acquisition of Reading
Ana Carolina Pena (Editor)
2009. ISBN: 978-1-60741-236-6

Approaches to Early Childhood and Elementary Education
Francis Wardle
2009. ISBN: 978-1-60741-643-2

Academic Administration: A Quest for Better Management and Leadership in Higher Education
Sheying Chen (Editor)
2009. ISBN: 978-1-60741-732-3

Recent Trends in Education
Borislav Kuzmanović and Adelina Cuevas (Editors)
2009. ISBN: 978-1-60741-795-8

Expanding Teaching and Learning Horizons in Economic Education
Franklin G. Mixon, Jr. and Richard J. Cebula
2009. ISBN: 978-1-60741-971-6

New Research in Education: Adult, Medical and Vocational
Edmondo Balistrieri and Giustino DeNino (Editors)
2009. ISBN: 978-1-60741-873-3

Disadvantaged Students and Crisis in Faith-Based Urban Schools
Thomas G. Wilson
2010. ISBN: 978-1-60741-535-0

Delving into Diversity: An International Exploration of Issues of Diversity in Education
Vanessa Green and Sue Cherrington (Editors)
2010. ISBN: 978-1-60876-361-0

Developments in Higher Education
Mary Lee Albertson (Editor)
2010. ISBN: 978-1-60876-113-5

The Process of Change in Education: Moving from Descriptive to Prescriptive Research
Baruch Offir
2010. ISBN: 978-1-60741-451-3

Career Development
Hjalmar Ohlsson and Hanne Borg (Editors)
2010. ISBN: 978-1-60741-464-3

Adopting Blended Learning for Collaborative Work in Higher Education
Alan Hogarth
2010. ISBN: 978-1-60876-260-6

Special Education in the 21st Century
MaryAnn T. Burton (Editor)
2010. ISBN: 978-1-60741-556-5

Challenges of Quality Education in Sub-Saharan African Countries
Daniel Namusonge Sifuna and Nobuhide Sawamura
2010. ISBN: 978-1-60741-509-1

Collaborative Learning: Methodology, Types of Interactions and Techniques
Edda Luzzatto and Giordano DiMarco (Editors)
2010. ISBN: 978-1-60876-076-3

Handbook of Lifelong Learning Developments
Margaret P. Caltone (Editor)
2010. ISBN: 978-1-60876-177-7

Virtual Worlds: Controversies at the Frontier of Education
Kieron Sheehy, Rebecca Ferguson and Gill Clough (Editors)
2010. ISBN: 978-1-60876-261-3

Health Education: Challenges, Issues and Impact
André Fortier and Sophie Turcotte (Editors)
2010. ISBN: 978-1-60876-568-3

Reading in 2010: A Comprehensive Review of a Changing Field
Michael F. Shaughnessy (Editor)
2010. ISBN: 978-1-60876-659-8

Becoming an Innovative Teacher Educator: Designing and Developing a Successful Hybrid Course
Qiuyun Lin
2010. ISBN: 978-1-60876-465-5

Sexuality Education
Kelly N. Stanton
2010. ISBN: 978-1-60876-465-5

Sexuality Education and Attitudes
Jovan Stanovic and Milo Lalic (Editors)
2010. ISBN: 978-1-60741-662-3

What Content-Area Teachers Should Know About Adolescent Literacy
National Institute for Literacy
2010. ISBN: 978-1-60741-137-6

Educational Change
Aden D. Henshall and Bruce C. Fontanez (Editors)
2010. ISBN: 978-1-60876-389-4

Music Education
João Hermida and Mariana Ferreo (Editors)
2010. ISBN: 978-1-60876-655-0

Computer-Assisted Teaching: New Developments
Brayden A. Morris and George M. Ferguson (Editors)
2010. ISBN: 978-1-60876-855-4

EDUCATION IN A COMPETITIVE AND GLOBALIZING WORLD SERIES

COMPUTER-ASSISTED TEACHING: NEW DEVELOPMENTS

BRAYDEN A. MORRIS
AND GEORGE M. FERGUSON
EDITORS

Nova Science Publishers, Inc.
New York

For permission to use material from this book please contact us:
Telephone 631-231-7269; Fax 631-231-8175
Web Site: http://www.novapublishers.com

LIBRARY OF CONGRESS CATALOGING-IN-PUBLICATION DATA

Available upon request

ISBN: 978-1-60876-855-4

Published by Nova Science Publishers, Inc. ✣ New York

CONTENTS

Preface		**xi**
Chapter 1	**Computer-Assisted Teaching on Science Education: A Student and Teacher Perspective** *Maria Limniou, Neil A. Burton and J. Christopher Whitehead*	**1**
Chapter 2	**Computer-Assisted Teaching and Concept Learning in Science: the Importance of Designing Resources from a Pedagogic Model** *Keith S. Taber*	**37**
Chapter 3	**Inquiry web-Based Learning to Enhance Knowledge Construction in Science: a Study in Secondary Education** *Manoli Pifarré Turmo*	**63**
Chapter 4	**Using Computer-Assisted Teaching to Promote Constructivist Practices in Teacher Education** *Rebekah K. Nix and Barry J. Fraser*	**93**
Chapter 5	**Implications of Expertise Reversal Effect for Multimedia Learning** *Slava Kalyuga*	**117**
Chapter 6	**Comparison of Multimedia Computer-Assisted Instruction, Traditional Instruction and Combined Instruction on Knowledge Acquisition and Retention of Setting Skill in Volleyball** *Nicholas Vernadakis, Panagiotis Antoniou, Eleni Zetou, Maria Giannousi and Eftimis Kioumourtzoglou*	**133**
Chapter 7	**New Forms of and Tools for Cooperative Learning with Social Software in Higher Education** *Sandra Schaffert and Martin Ebner*	**151**

Chapter 8	**Computer-Assisted Teaching and New Technologies** *D. Mutter and J. Marescaux*	**167**
Chapter 9	**Expert Commentary: E-Technology Use and Abuse in University Classrooms** *Rocci Luppicini*	**179**
Chapter 10	**Participants in the Community of Knowledge: Experiences with a Wikipedia Writing Assignment for Students** *Normann Witzleb*	**191**
Chapter 11	**Affordances of "My Learning Space" for Constructivist Learning** *Nadarajah Loganathan and Qiyun Wang*	**201**
Index		**213**

PREFACE

Computer-assisted teaching and the learning environments that support it have risen rapidly at the school and college level over recent years. Computer-assisted teaching usually includes multimedia instructional formats, such as animation, video and simulation, and virtual learning environment such as WebCT and Blackboard. In this book, the views of chemistry students on the use of computer-assisted teaching is demonstrated. How the teacher can effectively use his electronic material resources into blending learning activities are shown, which involve a combination of face-to-face interactions and technologically mediated interactions between students and teachers. This book also considers the challenge of producing materials to support conceptual learning in science, and the type of pedagogic models that are needed for successful computer-assisted teaching of science. Moreover, the authors report how the incorporation of information and communications technology (ICT) into a teacher professional development program fostered constructivist teaching/learning practices in school classrooms. In addition, expertise reversal effect relates to the interactions between levels of learner prior knowledge and effectiveness of different instructional techniques and procedures. This book reviews recent empirical findings associated with the expertise reversal effect in multimedia and hypermedia learning environments, their interpretation within a cognitive load framework and implications for the design of learner-tailored multimedia.

Chapter 1 - Computer-assisted teaching and the learning environments that support it, have risen rapidly in the agenda of Higher Education over the recent years. In various educational institutions nowadays more and more teachers make an effort to adopt information technology to their tutorials, assessment methods and laboratory sessions in an attempt to improve students' motivation and performance. This integration facilitates the teachers to prepare quite flexible teaching approaches and meet the diverse students' learning needs and expectations. Computer-assisted teaching usually includes multimedia instructional formats, such as animation, video and simulation, and/or virtual learning environments such as WebCT and Blackboard. Parameters which influence the effectiveness of computer-assisted teaching are the design and the integration of the electronic materials into courses, the learners' characteristics, their needs and expectations for the courses and the teachers' characteristics in terms of web-based professional development. The aim of this chapter is to demonstrate the views of chemistry students on the use of computer-assisted teaching and teachers' efforts to incorporate blended learning in their lectures. By purposely choosing examples from science education, which combines lectures and practical work, the authors describe the students' opinion on computer-assisted teaching and then use this feedback to demonstrate how teachers can build a successful computer-assisted course. The authors further examine the correlation between the electronic learning material and the students' background knowledge and the characteristics that allow them to adapt it into their learning style. Finally, the authors show how the teacher can effectively use his electronic material resources into blending learning activities, which involve a combination of face-to-face interactions and technologically mediated interactions between students and teachers.

Chapter 2 - Computers and related information and communication technologies (ICT) are increasingly being employed in teaching at school and college level. Like any new educational technology, computers have strengths and limitations; and it is important that computers are used to support existing educational aims when they are appropriate tools, rather than simply being adopted to do what they are good at because ICT is seen as intrinsically 'good'. In science teaching, computers offer a number of useful properties, both related to their use in data collection and analysis, and their ability to offer high quality simulations. These modes of use offer much to support the classroom teacher. However, computers are increasingly being seen as suitable tools for 'delivering learning' in individual study as a supplement to, or even an alternative to, the teacher. Whilst this offers more flexibility - in where and when studying can occur - resources intended to 'teach' learners when the teacher is not present are a different proposition to resources provided as tools for the teacher to employ flexibly to support their own teaching. Conceptual learning in science is well recognised as often being problematic even with motivated students and skilled teachers. This chapter considers the challenge of producing materials to support conceptual learning in science, and the type of pedagogic models that are needed for successful computer-assisted teaching of science. As an example, the chapter discusses the pedagogic model employed in developing learning material in physics for use in further education colleges in England, and the responses of students to the resources.

Chapter 3 - Science is a complex topic and over the past two decades information and communication technologies have had an advantageous impact on science teaching and learning. When information and communication technologies are used into classrooms, students play an active role, learn at school and outside school, improve their teamwork skills, ask more questions, find answers to questions and show a higher interest in learning. The development of pedagogical strategies suited to classroom use of online resources should be an important priority for the science education community.

Although many studies have demonstrated the effectiveness of computer-based instruction in the students' achievement in science, others suggest that students get lost on the Internet when searching information. However, other works have reported that students have difficulties in regulating their learning when using hypermedia environments. Science educators claim the necessity of attempting to facilitate students' learning of science topics by using scaffolds, or instructional aids, designed to develop appropriate web searching and managerial strategies and to support students' web-based learning.

In response to this educational necessity, the author's work offers more insight into the challenge of combining information and communication technology with Inquiry Web-Based Learning and procedural guidance in order to increase the students' involvement and responsibility for their own learning of complex science topics. The main goal of the present study is to integrate technology and inquiry into science classrooms to assist students build their own knowledge and answer meaningful driving questions.

A challenging instructional approach was designed, implemented, and evaluated with 127 secondary education students. Specific scaffolds were provided to guide students through complex tasks, to help them develop scientific knowledge, and to provide support to enable them to transfer what they learnt during the web-based activity. All activities were designed keeping in mind the WebQuest structure but also including the circular notion of inquiry process presented by Lim (2004). According to Lim (2004) an inquiry process has the following elements: Ask, Plan, Know, Explore, Construct, and Reflect.

Quantitative assessments showed that students in the experimental group outperformed the ones in the control group in the post test, indicating the positive effect of the instructional design. Some clues to use adaptive scaffolds effectively were exposed. Some of these scaffolds are tables, simulators, applets, visualizations, conceptual maps, and note-taking areas/boxes.

This research lends evidence to questions regarding the value of students engaging in on-line inquiry learning to enhance science content understanding. The conclusions drawn in this work present a positive and promising path in the design of instructional processes capable to enhance students' digital literacy and, consequently improve student's content understanding using web information.

Chapter 4 - In this chapter, the authors report how the incorporation of information and communications technology (ICT) into a teacher professional development program fostered constructivist teaching/learning practices in school classrooms. Changing the teachers' learning environment at the university level, based partly on computer-assisted teaching, fostered similar changes in their students' middle-school learning environments. The Constructivist Learning Environment Survey (CLES), which provided a critical scaffold for design and delivery, was administered to 845 students of 17 teachers. When compared to other classrooms within the same middle schools, these science classrooms were perceived to be significantly more positive in terms of the CLES scales of Personal Relevance, Shared Control, Critical Voice, and Student Negotiation. Qualitative data from observations and in-depth case studies provided further insight into the ways in which the ICT-enhanced program changed the classroom practice of experienced science teachers. Four assertions regarding the implementation of constructivism were identified relating to teacher collaboration, efficacy, understanding, and facility.

Chapter 5 - Expertise reversal effect relates to the interactions between levels of learner prior knowledge and effectiveness of different instructional techniques and procedures. Designs and techniques that are effective with low-knowledge learners can lose their effectiveness and even have negative consequences for more proficient learners. This paper reviews recent empirical findings associated with the expertise reversal effect in multimedia and hypermedia learning environments, their interpretation within a cognitive load framework, and implications for the design of learner-tailored multimedia.

Chapter 6 - The purpose of this study was to examine the effect of multimedia computer - assisted instruction (MCAI), traditional instruction (TI), and combined instruction (CI) methods on learning the skill of setting in Volleyball. Additionally, a comparison of the students' attitudes towards the MCAI and TI methods was made. Forty-eight middle school students of seventh and eighth grade were randomly assigned into three teaching method groups: TI, MCAI and CI. Each group received ten 45-min periods of instruction divided into 3 sections: a) 5-min introduction, b) 30-min instructional time and c) 10-min questions and review. Students took pre-, post-, and retention written test covering rules and strategy of the games. Participants in the CI group also completed a post-test attitude survey towards the MCAI and TI methods. Two-way analysis of variances (ANOVA), with repeated measures on the last factor, were conducted to determine effect of method groups (MCAI, TI, CI) and measures (pre-test, post-test, re-test) on knowledge test. Paired samples t-test analyses were conducted to measure students' attitude towards the MCAI and TI methods. Post-test results indicated no significant differences between the groups concerning the written test. Nevertheless, the attitude test scores of the CI group were more favourable to MCAI method than the TI method. Retention test results showed that groups retained the knowledge acquisition. However, the combine method of instruction tended to be the most effective on cognitive learning.

Chapter 7 - Since the new generation of Internet technology, called Web 2.0, has been introduced, a change of how users are dealing with the World Wide Web has been get into going. If access to the Web is available, today nearly anyone can actively participate and communicate online. Of course this recent evolution of the Web influences also the field of education. Former e-learning was mainly characterized by the use of content offered within learning management systems. Nowadays so called "Social Software" enables new possibilities and didactical approaches. In this chapter the authors give a short overview of how Social Software can support cooperative learning and how new technologies can enhance higher education in a meaningful new way. After a short introduction to the basics of cooperative learning different Social Software applications are classified and described. Practical examples are presented to show the general usage. In the end the authors conclude that these technologies have great impact on teaching and learning, as it will help to enhance education at universities.

Chapter 8 - Many procedures in a broad spectrum of clinical specialities can be managed through a minimally invasive surgical (MIS) approach. Indeed, minimal access surgery has become increasingly accepted by the surgical community, but the surgeon is faced with the necessity of following a specific training.

Over the past two decades, advances in computer-assisted technology and telecommunications have deeply altered surgical education. The role of the traditional "Professor of Surgery" has been superseded by the quantity and quality of knowledge available not only in the media and on the Internet, but also in CD-ROMs, DVDs, stream videos, webcasts and telementoring. Nowadays, information and knowledge is readily accessible for all to find. What is more, the knowledge of highly skilled teachers is available worldwide thanks to the Internet and other multimedia technologies.

As new perspectives are opening up on the surgical horizon, such as natural orifice transluminal endoscopic surgery (NOTES), the need to apply new technologies to teaching and training as well as in daily clinical practice has become evident. This encapsulates multimedia, computer-assisted learning, videoconferencing, virtual reality and computer-assisted simulation in order to maximize surgical education. Telementoring through videoconferencing renders the acquisition of the cognitive aspects of surgical procedures easier through the observation of experts during hands-on practice. The development of computer technology allows to recreate a virtual 3D model of the patient, capable of simulating the patient's anatomy on a computer; computerized 3D images also allow virtual navigation around and into the anatomical structures, simulation of various interventions as well as automated volume calculation. In addition, a virtual reconstruction of the patient may also enhance diagnosis and enable a preoperative simulation and a preoperative training.

Finally, the computerized data will allow the scheduling of automated surgical procedures. Robotics have demonstrated their ability in performing long-distance, remote-controlled surgical procedures. Surgeons require access to information whenever and wherever, and the Internet offers them that very chance, along with observation, training and education. Websites dedicated to surgery such as www.websurg.com incorporate all aspects of surgical training including fully illustrated and animated operative techniques, videos and experts' recommendations that, with the continuous updating of information, represent the future tool for the transfer of surgical and scientific knowledge. Using all these technologies, education may be improved through WeBSurg's virtual university as it gathers academic knowledge, education and expertise from around the world without any constraints of time and/or space. It will also facilitate the standardization of surgical treatments.

Chapter 9 - This expert commentary addresses academic and non-academic use of electronic technology within undergraduate university classrooms. Despite the potential benefits of e-technologies, there are substantial obstacles to effective and responsible e-technology use in university classrooms that can given rise to heated debates between university administrators, instructors, and students concerning the appropriate use of e-technologies within undergraduate university classes. Based on the author's teaching and research experience working with educational technologies in Higher Education, this chapter identifies a number of serious problems and concerns revolving around current e-technology practices in university classes. This expert commentary draws on professional experience and research to provide insight into the use and abuse of e-technologies within undergraduate university classrooms. It also offers recommendations on how to leverage the educational use of e-technology within undergraduate education.

Chapter 10 - Improving students' computer literacy and instilling a critical approach to internet resources are important educational aims for today's generation of students. This Commentary will examine how a writing exercise in the style of a Wikipedia article can be used to facilitate these skills. Students in an elective unit in Comparative Law were asked to write, and review, an article on an issue, concept or scholar in Comparative Legal Studies. The Commentary will describe the rationale for adopting this form of assessment, how it was integrated into the teaching and assessment structure of the unit, and how students responded to the exercise. It will critically evaluate this novel form of assessment. The aim of the Commentary is to provide some practical guidance on the potential benefits of this form of assessment and when it might be usefully employed.

Chapter 11 - Information and communication technology (ICT) has been popularly used not only in schools but also in various organizations like training centers. They do not just simply use ICT tools to present content or let students or trainees do drill-and-practice. They are now more using an integrated learning management system to conduct and facilitate the whole teaching or training process. In this chapter, the authors present the basic features and pedagogical, social, and technological affordances of a learning management system (called my learning space) which is used by the Signal Institute of the Singapore Armed Forces (SAF). The perceptions of trainees on the use of the learning space are also briefly descried.

In: Computer-Assisted Teaching: New Developments
Editors: B. A. Morris, G. M. Ferguson, pp. 1-36
ISBN: 978-1-60876-855-4

Chapter 1

COMPUTER-ASSISTED TEACHING ON SCIENCE EDUCATION: A STUDENT AND TEACHER PERSPECTIVE

Maria Limniou**, *Neil A. Burton and J. Christopher Whitehead
The University of Manchester, UK.

ABSTRACT

Computer-assisted teaching and the learning environments that support it, have risen rapidly in the agenda of Higher Education over the recent years. In various educational institutions nowadays more and more teachers make an effort to adopt information technology to their tutorials, assessment methods and laboratory sessions in an attempt to improve students' motivation and performance. This integration facilitates the teachers to prepare quite flexible teaching approaches and meet the diverse students' learning needs and expectations. Computer-assisted teaching usually includes multimedia instructional formats, such as animation, video and simulation, and/or virtual learning environments such as WebCT and Blackboard. Parameters which influence the effectiveness of computer-assisted teaching are the design and the integration of the electronic materials into courses, the learners' characteristics, their needs and expectations for the courses and the teachers' characteristics in terms of web-based professional development. The aim of this chapter is to demonstrate the views of chemistry students on the use of computer-assisted teaching and teachers' efforts to incorporate blended learning in their lectures. By purposely choosing examples from science education, which combines lectures and practical work, we describe the students' opinion on computer-assisted teaching and then use this feedback to demonstrate how teachers can build a successful computer-assisted course. We further examine the correlation between the electronic learning material and the students' background knowledge and the characteristics that allow them to adapt it into their learning style. Finally, we show how the teacher can effectively use his electronic material resources into blending learning activities, which involve a combination of face-to-face interactions and technologically mediated interactions between students and teachers.

* Corresponding author: E-mail: Maria.Limniou@manchester.ac.uk

INTRODUCTION

In science education, the curriculum consists mainly of lectures and laboratory sessions. Ideally, through lectures, the students can understand the theory, which they then can put into practice during the laboratory sessions. The most popular teaching approach is the traditional or conventional way of teaching. Following that approach, students are taught in a monologue lecture environment. The material is delivered as predetermined texts and the learning environment is usually passive, giving little motivation to students to participate actively in the class. Usually, students get lost in the process of taking lecture notes and combining the technical and scientific concepts with their practical application. As a result the students seldom have the opportunity for critical thinking during the class and they cannot easily adopt the scientific way of thinking (Seng & Mohamad, 2002). On the other hand, by following the traditional way of teaching in the laboratory, the teachers can effortlessly predetermine the activity's outcome, as the students perform experiments following the manual mechanically. The laboratory activities are usually performed by a large number of students with minimal involvement from the teacher and often, due to lack of funds, with minimal resources, space, equipment and technical support (Domin, 1999).

What often teachers fail to realise is that during the laboratory session, students receive a huge amount of information, such as the recognition of equipment and the associated handling, instrumentation and safety requirements in a laboratory environment, all in a short period of time (Johnstone, 1997). As a result, students often struggle to absorb all this information at the same time and cannot respond to the demands of the practical work at hand. Ideally, learners should receive the new information gradually in order to process and absorb into it their working memory and also have a chance to connect it with their previous knowledge, which is stored in their long-term memory. The working memory has space for only a limited number of pieces of information at the same time. This process is described by the cognitive load theory and the model of the information process for learning (Figure 1) (Gabel, 1999; Bailey & Garratt, 2002).

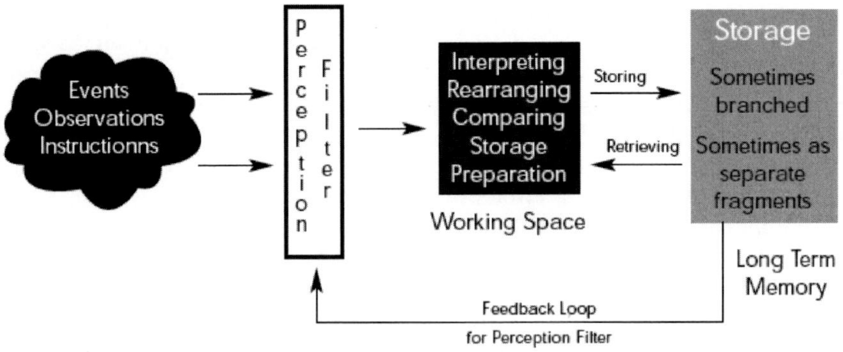

Figure 1. Information processing model (after Gabel, 1999).

If students receive huge amounts of information at the same time, then their working memory gets overloaded and the learning process cannot continue successfully (Johnstone, 1997). Additionally, students who lack a solid theoretical background will face additional difficulties grasping the aim of the experiment they are about to perform. In essence, the conventional teaching approach is the most criticized as placing little emphasis on thinking (Gunstone & Champagne, 1990), being an ineffective means of conceptual change (Merritt et al., 1993) and being unrealistic in the simulation of scientific experimentation (Domin, 1999).

The majority of science teachers in an effort to tackle the aforementioned difficulties in the laboratory session, give a short lecture presentation at the beginning of the laboratory session in an attempt to emphasize the experimental procedures and the experimental data collection (pre-laboratory activities) (Nicholls, 1999). However, the learning environment remains passive, as the teacher presents technical and scientific concepts and principles on the blackboard. Other teachers tend to summarise the experimental procedure by making a combination between the theory and the experimental data at the end of the laboratory session (post-laboratory activity) (Garratt, 1997; Nicholls, 1998). During this session, the teacher explains to students the rationale of what they have done during the laboratory, linking theory with practical work. Also, the teacher gives instructions to students on how to process and analyse the experimental data. Following that teaching approach, students can consolidate the theory with laboratory activity using their previous knowledge.

Learning process is influenced by students' previous knowledge and it is dependent on their misconceptions or their well constructed knowledge in their long-term memory (Bodner, 1986; Ausubel, 2000; Bodner et al., 2001). Their lack of previous knowledge in combination with their overloaded working memory often hinders students' attempts to make connections between the laboratory activities and the associated theory or to call from their long-term memory methods or techniques for processing and analysing data. Moreover, the limited time in traditional laboratory sessions or lectures does not allow teachers to clarify students' misconceptions or to provide students with the opportunity to refresh their previous knowledge. Therefore, students usually follow instructions from the laboratory manual mechanically, line-by-line, as they perform their experiments and thus miss the essence of the experiment (Garratt, 1997). According to the USA National Research Council (1996) students should participate actively in learning the scientific way of thinking which includes observations, hypothesis, law applications, collection, processing, analysis and interpretation of experimental data. However, in the conventional teaching approach the learning environment is passive and students do not have any active participation in the learning process. Furthermore, science education is comprised of difficult concepts and as teachers follow the conventional teaching approach they do not allow students to construct their own cognitive structure through active participation in the learning process. Teachers often fail to examine their students' previous knowledge and plan a lecture with the belief that students have all the same level as dictated from their school or faculty. This results in students becoming unsynchronised with the lectures; they often feel lost during their studies, become bored and inattentive in class, have a poor performance and in many cases change course. Further adding to the problem, teachers are unjustly on the receiving end of the students' frustration and either they may become overly critical of their students or they are embarrassed by their criticism (Felder & Silverman, 1988).

Another factor that influences the learning is the students' learning styles. According to Felder and Silverman (1988) learners can be divided into categories, such as active versus reflective learners, visual versus verbal learners, sequential versus global learners and sensing versus intuitive learners and teachers should follow different teaching approaches in order for their teaching to be characterised as effective. The student population has grown in Higher

Education and flexibility in teaching (in terms of time, pace, place and accessibly) becomes more and more a requirement. However, one realizes that *it is not possible to discover everything that affects what an individual student learns in a class, and even if instructors could, they would not be able to figure out the optimum teaching style for that student—the task would be far too complex. Moreover, even if a teacher knew the optimum teaching styles for all students in a class, it would be impossible to implement them simultaneously in a class of more than two students* (Felder & Brent, 2005). Thus, students who have different needs and different learning styles are not equally satisfied with the teaching approach and the learning environments.

The combination of Information and Communication Technology (ICT) with teaching and learning has risen rapidly up the agenda of Higher Education in recent years and it has proven a very useful tool in the teacher's never ending attempt to develop flexible teaching approaches in order to meet the diverse learning needs of their students. Computer-based applications provide useful teaching features (Clow, 1998) such as

- Hypertext, where highlighted words belong to text giving students the opportunity to collect information depending on their learning style and their needs.
- Computer-based multimedia, where a combination of visual and verbal information such as graphics and animations illustrates difficult concepts or phenomena to students.
- Formative computer marked assessment, where feedback is provided to students easily and quickly, giving them the opportunity to check their responses to questions and study the logical steps of the solution to a given problem.
- Computer-mediated communications, where (synchronous or asynchronous) communication between teacher(s) and student(s) and between students are enhanced through computer by using communication or discussion tools such as e-mail or messenger technology.
- Laboratory instrumentation, where students send commands to a real instrument or receive from it experimental data through a suitable computer interface (computer-controlled instruments/remote-controlled instruments).
- Virtual simulation, where the reality (examples from real world, instruments and experimental techniques) is represented computer desktop.

All of the above instructional computer applications features are involved in Computer-Assisted Teaching. For example, computer-based multimedia learning environments have been very promising in Higher Education, as they have the ability to combine text, sound and images and create an instructional environment which can be used by a great number of students simultaneously. Furthermore, teachers can use these computer-based applications and integrate them into different study modes (lectures, pre-laboratory or post-laboratory sessions) and/or use them in different ways such as presentation tools or educational material for students' assessment.

The three most important computer-based applications which belong to the board area of multimedia are:

1. Animations: which *are powerful medium for displaying how dynamic systems function in a space and time scale accessible to human perception* (Betrancourt et al., 2008). By using animations, the students' attention is focused on important aspects of a display, conveying procedural knowledge, demonstrating the dynamics of a topic and/or allowing exploratory learning through the manipulation of an object. Visualizations and animations are especially helpful in science education, as animations present information that changes with time by movement of one or more objects in the animations changing properties of objects (transformation), movements of objects on the screen (translation) and/or appearance of objects (transition) (Lowe, 2003; 2004). Animations allow learners to better process and absorb information in comparison to static pictures. For example, manipulation through animation allows learners to generate and display a large number of static pictures showing different states under different perspectives (Schnotz & Rasch, 2005).
2. Digital video: where visual and verbal information are electronically captured, recorded, processed, stored, transmitted, and reconstructed as a sequence of still images representing scenes in motion. It is a powerful medium that can provide narrative visualization and can engage multiple senses simultaneously creating a sense of *social presence* (Gunawardena, 1995). Digital video permits more effective interactivity and control, as video elements can be quickly selected by the user, or controlled by a computer program, in any desired sequence (Palmer, 2007). Presentations may involve audio and video of a lecture, along with a visual presentation of the presenter's slides (Homer et al., 2008).
3. Virtual Simulation: which mimic important elements of science phenomena, instruments or experimental procedures. By using computer simulations, students have the opportunity to understand unobservable phenomena in science and/or to familiarise themselves with laboratory techniques manipulating input values of variables which describe the system, observe their effects on output displays and manipulate the simulated instrument as they would a real one. Furthermore, the use of a simulator reduces the purchase and maintenance costs of the laboratory equipment, while enabling students to perform a variety of experiments in data collection and analysis (Thomas & Neilson, 1995; de Jong et al., 1999). In simulation-based learning students gradually infer the features of the concept model whilst they proceed through the simulation, which may lead to changes in their original concept (de Jong & van Joolingen, 1998). According to Chang et al. (2008), simulation-based learning is particular beneficial as it:

    1. Provides background knowledge
    2. Helps learners to make hypothese
    3. Helps learners to conduct experiments
    4. Helps learners to interpret data
    5. Helps learners to regulate the learning process

In Computer-Assisted Teaching, teachers use computer-based applications in different periods, places and ways in order to facilitate learning, as they provide different learning venues to students to better comprehend the course material. At this point we need to make clear; however that the Information and Communication Technology applications, which are used, are only tools for learning facilitation and teachers are not meant to be replaced in any way. Thus, teachers are responsible for finding and/or creating learning experiences, accessing students' previous knowledge through explanation of the real world and using compare and contrast techniques, co-operative and/or collaborative learning and other methods that will assist students to integrate new knowledge into their cognitive frameworks (Bunce, 2001). Moreover, the teacher who follows the traditional way of teaching can enhance the procedure by using computer-based applications, for example by uploading the course and/or onto supplementary material on a web space. Students can follow the teacher during the lecture without scribbling down notes from the lecture and read the material in their own time and space after the lecture. Additionally, the students who miss a lecture can prepare for the next one without leaving any knowledge gaps between the lectures. Also, through computers, collaboration activities can be facilitated and the teachers can establish discussions so that experiences can be shared between the teacher and the students making learning a more active process.

In order for computer-based applications to become effective educational tools for computer-assisted teaching, the context, the design, the integration of digital material into lectures or laboratory sessions and the learners' preferences on the delivered material should be taken into account. Thus, the challenge for teachers lies in identifying the conditions under which different applications are most effective for teaching and learning (Mayer & Moreno, 2002). Thus, for example, according to Tversky et al. (2002), the two conditions under which dynamic visualisations may be more effective than static are:

1. the structure and the content of a dynamic visualisation should be readily perceivable and comprehensible
2. the structure of a dynamic visualisation should correspond with the way that people conceive the process and the procedures that are visualised.

Cognitive load theory in its combination with the multimedia design explains the conditions which influence the effectiveness of computer-based applications and describes the general conditions which are more effective for computer-assisted teaching. Specifically, cognitive load theory describes the demands on working memory storage and the procedure of information processing in order for learners to store permanently in their long-term memory. The human cognitive architecture includes different subsystems such as sensors, working memory, and long term memory in order to receive, process, and store information. The central feature of this theory is the limited working memory capacity. Within cognitive load, a distinction was made between the nature of the learning task (intrinsic load), the format of instruction (extraneous load) and the development of cognitive schemata (germane load) (Chandler & Sweller, 1991; Sweller et al., 1998; Schnotz & Kürschner, 2007).

- Intrinsic load: is inherent to the material being learned. The natural complexity of information corresponds to the required working memory capacity. Therefore, the more complex the information is, the greater the intrinsic load is.
- Extraneous load: is caused entirely by the format of the instruction. The design, presentation and organisation of the learning material impose an extra mental effort on the learners' working memory. Extraneous load interferes with learning and, thus,

should be reduced as far as possible by eliminating irrelevant cognitive activities. The instructional design should aim to decrease extraneous load.

- Germane load: is caused by effortful learning resulting in schema construction and schema automation. In other words, germane load is the learners' mental effort to process the new information and to integrate it into existing knowledge schemata.

Unused working memory capacity because of low intrinsic load and low extraneous load may encourage learners to engage in conscious cognitive processing which is directly relevant to the construction of cognitive schema (Schnotz & Kürschner, 2007). As a general principle, learning material should be designed to reduce extraneous load in order to allow for the greatest amount of mental resources to be dedicated to germane load, which means that learners' working memory capacity should be available for intrinsic load and germane load (Paas & Merriënboer, 1994). Germane load requires an extra amount of working memory capacity beyond the intrinsic load (or extraneous) and this extra capacity is used for learning.

Despite their usefulness and robustness Computer-Assisted Teaching tools do not on their own guarantee success. For example, computer-based applications such as animations may be too complex or too fast to be accurately perceived; continuous events are also often conceived as sequences of individual steps. In that case the cognitive benefits of animations in the understanding, for example, of dynamic mechanical systems included in technical or scientific documents remain disputable, as students have to share their attention between text and animation. They are often very hard to perceive and to process and pose strong perceptual, conceptual and working memory demands (Boucheix & Schneider, 2009).

Multimedia learning theory is based upon three types of cognitive load (intrinsic, extraneous and germane load) which influences the learning process when the working memory processes visual and verbal information (Sweller, 1994). The verbal information is presented in either visual (printed text) or auditory form. The way that cognitive load theory is applied on the design of multimedia learning environments are involved in all the computer-based applications such as animations and virtual simulations giving emphasis on the extraneous cognitive load reduction and the reduction of the working memory capacity for learning (Bannert, 2002).

Many researchers have studied the effectiveness of multimedia learning environments in conjunction with cognitive load theory (Mayer & Moreno, 2002; Rouet et al., 2008; Wouters et al., 2008). Robinson (2002) has summarised the basic aspects of cognitive theory which are necessary in order to design a multimedia application. According to Mayer & Moreno (2002, 2003) the human cognitive system consists of two distinct channels for representing and controlling knowledge: an auditory-verbal channel and a visual-pictorial channel (the dual-channel theory) (Figure 2). Working memory has two separate but interrelated memory channels for processing verbal and visual information. However, working memory has a limited capacity. The visual-pictorial channel can be overloaded if too many pictures (or other visual material) are presented at one time. Similarly, the auditory-verbal channel can also be overloaded. When one channel is overloaded, the processing ability of the other one is limited. The learners organise the incoming words into a coherent verbal representation and the images into visual representation and integrate these two types of representations with relevant prior knowledge (meaningful learning and substantial cognitive processing in two memory channels (verbal and visual)).

There are three levels of cognitive load processes which are involved in multimedia learning (Figure 2) (Mayer, 2001, Mayer & Moreno, 2002; 2003).

1. Selection: verbal and visual information are selected by two separate but interrelated memory channels-verbal and visual channels.
2. Organisation: learners organise incoming words into a coherent verbal representation and images into visual representation and
3. Integration: learners integrate these two types of representations with relevant prior knowledge.

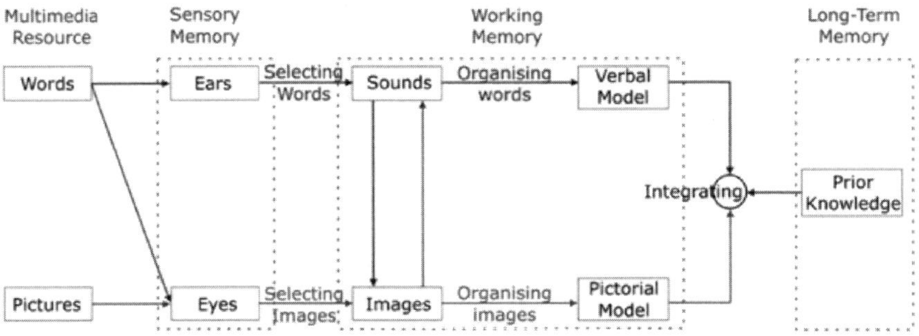

Figure 2. Mayer and Moreno's schema for the processing of knowledge from a multimedia resource (Mayer & Moreno, 2003).

However, a split-effect occurs when the learner's attention must be split between two (or more) sources of information that have to be processed simultaneously in order to be comprehensive A design guideline to overcome this split effect is verbal information to be provided as explanations rather than verbal information (Schnotz & Kürschner, 2007). However, the verbal channel is not always feasible (demand additional hardware, difficulty for deaf or hearing-impaired students, high demand on the auditory system) (Wouters et al., 2008). Additionally, *a modality effect occurs when multiple sources of information have to be integrated for comprehension and when using only visual sources of information would require learners to split their attention* (Schnotz & Kürschner, 2007). The limited cognitive capacity has to be effectively distributed between the written text and the pictorial information (Wouters et al., 2009). According to Tabbers et al. (2004) written explanations are more effective than spoken explanations when learners have the control of the computer-based application. Instructions which require a higher cognitive load for constructing a mental representation cause an extraneous cognitive load upon the learner's working memory.

Many researchers have studied under what conditions various media are effective educational tool for different learners. For example, *learners with lower prior knowledge used a text to a considerably lower extent when there were also pictures presented than when no pictures are available* (Schnotz, 2008). Additionally, according to Homer et al. (2008), *video increased the cognitive load for learners with a lesser preference for visual materials, but decreased the cognitive load for learners with greater preferences for visual materials.* According to Mayer & Moreno (2000) students learn better from an electronic material, when

1. text and images are integrated on the screen space (Spatial Contiguity)
2. text and images are integrated in time by providing information simultaneously or in small bites (Temporal Contiguity) and
3. information is presented concurrently or sequentially (Modality)

Computer-based applications such as animations or virtual simulations are used as educational tools to explain difficult scientific concepts. Teacher can visualise an internal part of a phenomenon or display internally an instrument's structure and students can easily work with different implementations on different examples. These activities can permit knowledge elicitation and understanding the transference of theoretical concepts to the real world can be earned out. The integration of computer-based applications into face-to-face or e-learning activities can promote the collaborative work and a better adaptation to the learning styles of students, for interactions among students and/or teacher.

Constructivism and Collaborative Learning explain the teacher's role as a knowledge facilitator and the role of collaboration in teaching and learning. According to constructivism, knowledge cannot be transferred from one person to another, but it is an active procedure where the learner constructs the knowledge through interactions with the environment (Bodner, 1986). Learning occurs when there is a change in the learner's existing ideas either by adding some new information or by reorganizing what is already known. The knowledge, which is constructed by the learner, is not a simple transmission of facts from the teacher to the students, but a continuous and active process on both sides. According to von Glasersfeld (1988) knowledge has a social construction where concepts are developed in a process of fine-tuning involving interaction with others. Constructivism and Collaborative Learning (CL) have established the social interdependence where students consolidate their knowledge by teaching one another. Collaborative Learning (CL) involves the mutual engagement of learners in a coordinated effort to solve a problem or to examine an issue all together. Basic points of Constructivism and Collaborative Learning (CL) are the following (Matthews et al., 1995; Kirschner, 2001):

- Learning takes place in an active mode
- The teacher is more a facilitator than "sage on the stage"
- Teaching and learning are shared experiences between teacher and student
- Students participate in small-group activities
- Students must take responsibility for their learning
- Discussing and articulating one's ideas enhances the ability to reflect on one's own assumptions and thought processes
- Students develop social and team skills through the give-and–take of consensus-building
- Students experience diversity, which is essential in multicultural democracy.

It is useful for educational purposes to focus not only on the types of multimedia formats and the way that they are designed but on the relationships between different modes of learning, such as face-to-face and e-learning approaches, for the enhancement of teaching approaches (blended learning). According to Bliuc et al. (2007), *blended learning describes activities that involve a systematic combination of co-present (face-to-face) interactions and technologically-mediated interactions between students, teachers and learning resources. The field is relatively new and many researches are still in exploratory mode: aiming to discover which are the more robust variables or constructs that explain successful experiences of face-to-face learning when it is combined with some technologically-supported learning.* In blended learning two environments are used: 1. the traditional face-to-face and 2. virtual learning environments (distributed environments). The latter are used as an expansion to the face-to-face learning. In the traditional face-to-face learning environment, the teaching is conducted in a synchronous environment, while by using Virtual Learning Environments the teaching is mainly focused asynchronously. Virtual Learning Environments (VLEs) such as WebCT, Blackboard and Moodle have been adopted as blended learning platforms by numerous universities in order for the teacher to have a flexible virtual learning environment to deliver on-line computer-marked assessment and/or course material in addition to standard classes. These are on-line management systems which provide a means of organising supported learning environments by allowing teachers to automate and provide information on selective portions of course materials and/or to give opportunities between the students and the course instructor to interact through collaboration (synchronous or asynchronous). Students have access to the Virtual Learning Environments and its material whenever and wherever they want, while possibilities for recording the time that the students spend in an on-line quiz, the number of students viewing a course and other useful statistical data are available to tutors (Blas & Serrano-Fernandez, 2009; Ngai et al., 2007; Bunce et al., 2006; O'Connor & Ross, 2004).

Delivery of feedback is another area that can greatly benefit from the use of Virtual Learning Environments in the teaching approach. Following the conventional approach, students often feel disappointed by the traditional delivery methods of feedback, where they have to wait for several weeks to receive comments on their performance in the form of written feedback from lecturers. In addition, feedback delivered to students in the traditional written form often fails to deliver on its primary objective, which is to clarify student's misconceptions and motivate them. To address this issue, the teacher can develop computer-marked formative assessments through Virtual Learning Environments, which allow students to get feedback quickly and easily. Two major categories of virtual learning environments are identified 1. virtual learning systems designed for use in the classroom setting (involving synchronous interactions) and 2. distributed virtual learning systems designed for environments in which the learners and instructions are distributed across time and/or geographic distance (involving synchronous and/or asynchronous interactions) (Alavi & Leidner, 2003).

Following methodologies for computer-assisted teaching encourages the learning increasing the experience, the active role and the social abilities of students in the building of the knowledge and promoting the role of the teacher/computer-based applications to a mediator. The proposed examples for laboratory instrumentation courses, lectures and the embedding of computer-based applications into Virtual Learning Environments demonstrate how the traditional teaching approach can be effectively enhanced by computer-assisted applications. However, before a description of computer-assisted teaching methodologies, a study is presented for students' preferences for different computer-based applications which support a computer assisted teaching approach.

STUDENTS' PERSPECTIVES FOR COMPUTER-BASED APPLICATIONS

Aim and Objectives

The aim of this section is to present the views of two groups of students (1st and 2nd year chemistry students) on the use of four different computer-based applications in terms of the students' preferences on these applications and the alternative forms of delivery (study modes). In order for these perspectives to be part of an objective investigation, the cognitive topic is the same for the four different applications. Students evaluated the four different applications regarding their effectiveness and usefulness on the learning and teaching, and classified them in combination with the different delivered study modes such as face-to-face and e-learning.

Description

The four applications are designed to support the same topic (acid-base titration). This technique is vital for all the first and second year chemical curricula and chemistry students attend various lectures and perform various experiments in the laboratory on this topic.

1st Application-animation without text support (AN1)

This animation presents to the students the titration technique without any verbal explanations. By using very clear graphics in this animation, students can observe the steps of the technique, process the visual information and integrate it into their existing schema of knowledge or they can produce a new one in their long term memory. The students who have misconceptions on the titration technique have time to process the various steps in the technique in order to clarify and consolidate it.

2nd Application-animation with text support (AN2)

In this animation, the students can find not only visual information but verbal information. The technique is presented step-by-step accompanied with appropriate labels which explain what is depicted in each frame of the animation. The students, who are not familiar with the technique, can observe the steps of the technique by reading the text and understand the reason why those steps are followed. In this way, the students construct a new schema for the acid-base titration or they can revise their already existing schema.

3rd Application-simulation which included simple tasks (SIM1)

By using this simulation the students perform simple virtual experiments choosing between three different specific reactions. After choosing an experiment, students can observe on the computer's desktop the experimental equipment and the procedure. The set up is the same as in the real laboratory environment. Following the simulation's instruction students perform the experiment interacting with simulation interface, observe the results of their actions on the computer desktop and make the appropriate calculations by using the simulation's experimental data. Thus, students participate actively in the learning process and they have time to process the new information in their working memory recalling their prior knowledge from their long-term memory.

4th Application-simulation which included theory and complicated tasks (SIM2)

This simulation is designed to introduce the students to acid-base titration by using a pH-meter and to train the students on acid-base titration curves (Papadopoulos & Limniou, 2003). In the virtual environment, dynamic and graphical information represent visually the laboratory technique and a large number of samples are available for students' training. The number of samples and the fact that students can set up their own experimental conditions make this simulation more flexible for learning. The experimental data is recorded on the virtual instruments and the simulation curves are plotted in the graphics panel. Students can save data as an ASCII file in order to import it into spreadsheets (Figure 3). Detailed theory and more complicated tasks than using SIM1 are included in the SIM2 simulation. Thus, students use virtual simulation to develop a complete cognitive schema regarding acid-base titration. They start to prepare samples, set up the experimental procedure virtually, get experimental data and interpret them as they can do in the real laboratory. Through that learning process, students generate and test hypotheses, make decisions, and explore and link the new information to their existing knowledge.

Results

This section presents a study which was conducted among two groups of chemistry students, twenty first year students (FYS) and twenty second year students (SYS). The 1st and 2nd year chemistry students classified the four different aforementioned computer-based educational formats into the following main categories:

1. general questions regarding the use of computer-based applications on education (familiarity with e-learning courses and computer-based applications),
2. interest and motivation; usefulness and effectiveness; and educational interface and student's learning style
3. computer-based educational format and integration into teaching procedure and
4. basic skills and blended learning

All responses provided were in the form of a five-point Likert-type scale (1=strongly disagree, 5=strongly agree) and we compared the students' responses by using t-test statistical analysis (independent samples test) (Table 1 and Table 2).

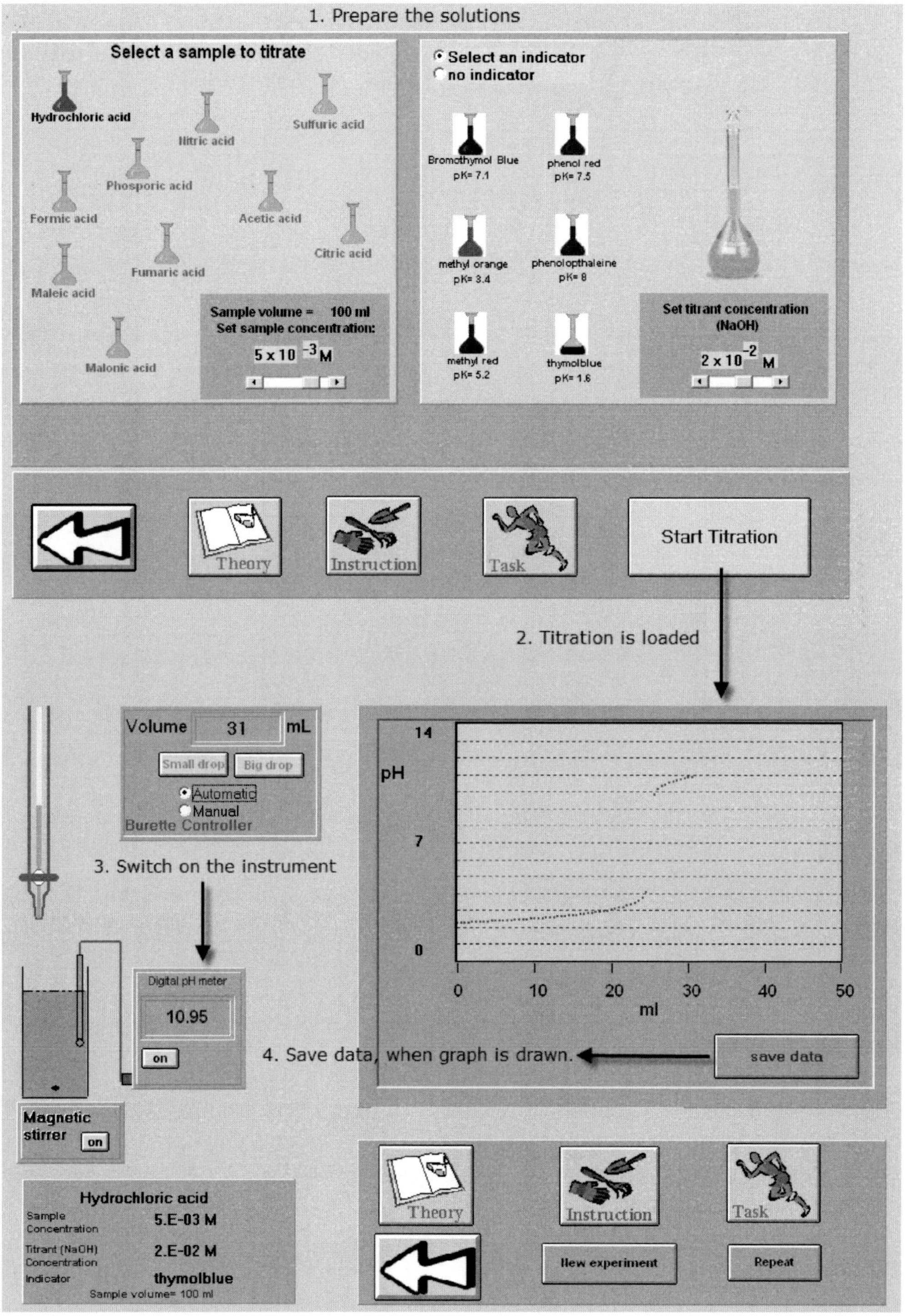

Figure 3. Main parts of the SIM2 simulation which illustrates the technique and the main function steps.

Table 1. t-test statistical* analysis for the AN1, AN2, SIM1 and SIM2 educational materials in order to compare the two group of students' responses on the six variable (interest, motivation, usefulness, effectiveness, interface, learning style).

	Interest	Motivation	Usefulness	Effectiveness	Clear interface	Learning style
AN1	FYS: *M*=2.6, *SD*=0.7 SYS: *M*=2.3, *SD*=0.7 *t*(38)=1.352, *MD*=0.300, *p*=0.184	FYS: *M*=2.4, *SD*=0.9 SYS: *M*=2.0, *SD*=0.7 *t* (38)=1.566, *MD*=0.400, *p*=0.126	FYS: *M*=2.7, *SD*=0.7 SYS: *M*=3.0, *SD*=0.9 *t* (38)=-1.189, *MD*=0.300, *p*=0.184	FYS: *M*=2.6, *SD*=0.5 SYS: *M*=2.6, *SD*=0.6 *t*(38)=0.284, *MD*=0.050, *p*=0.778	FYS: *M*=3.4, *SD*=0.8 SYS: *M*=3.4, *SD*=0.7 *t* (38)=0.000, *MD*=0.000, *p*=1.000	FYS: *M*=2.6, *SD*=0.8 SYS: *M*=2.6, *SD*=0. 9 *t* (38)=-0.184, *MD*=-0.050, *p*=0.855
AN2	FYS: M=3.7, SD=0.9 SYS: M=2.5, SD=0.6 t(38)=4.910, MD=1.200, p=0.000	FYS: M=3.3, SD=1.1 SYS: M=2.2, SD=0.7 t(38)=3.917, MD=1.150, p=0.000	FYS: M=3.5, SD=0.7 SYS: M=3.6, SD=0.8 t(38)=-0.438, MD=-0.100, p=0.664	FYS: M=3.2, SD=1.1 SYS: M=2.3, SD=0.7 t(38)=3.226, MD=0.950, p=0.003	FYS: *M*=3.7, *SD*=0.6 SYS: *M*=3.5, *SD*=0.8 *t* (38)=0.940, *MD*=0.200, *p*=0.353	FYS: M=3.7, SD=0.7 SYS: M=2.1, SD=0.7 t(38)=6.633, MD=1.550, p=0.000
SIM1	FYS: M=3.6, SD=1.0 SYS: M=3.9, SD=0.9 t(38)=-0.982, MD=-0.300, p=0.333	FYS: M=3.9, SD=0.9 SYS: M=4.0, SD=0.9 t(38)=-0.529, MD=-0.150, p=0.600	FYS: *M*=3.7, *SD*=0.6 SYS: *M*=3.8, *SD*=0.6 *t* (38)=-0.533, *MD*=-0.100, *p*=0.597	FYS: *M*=3.6, *SD*=1.1 SYS: *M*=3.2, *SD*=1.2 *t* (38)=0.927, *MD*=0.350, *p*=0.360	FYS: *M*=3.5, *SD*=0.6 SYS: *M*=3.5, *SD*=1.0 *t* (38)=0.191, *MD*=0.050, *p*=0.849	FYS: *M*=3.9, *SD*=0.6 SYS: *M*=3.7, *SD*=0.6 *t* (38)=0.819, *MD*=0.150, *p*=0.418
SIM2	FYS: *M*=3.1, *SD*=0.8 SYS: *M*=3.9, *SD*=0.6 *t* (38)=-3.241, *MD*=-0.750, *p*=0.02	FYS: *M*=3.2, *SD*=0.8 SYS: *M*=3.7, *SD*=0,9 *t* (38)=-1.993, *MD*=-0.550, *p*=0.054	FYS: *M*=3.5, *SD*=0.7 SYS: *M*=3.9, *SD*=0.8 *t* (38)=-2.026, *MD*=-0.450, *p*=0.050	FYS: *M*=3.0, *SD*=1.0 SYS: *M*=3.9, *SD*=1.0 *t*(38)=-2.733, *MD*=-0.850, *p*=0.009	FYS: *M*=3.1, *SD*=0.7 SYS: *M*=3.9, *SD*=0.8 *t* (38)=-3.474, *MD*=-0.850, *p*=0.001	FYS: *M*=3.0, *SD*=1.0 SYS: *M*=4.1, *SD*=0.6 *t*(38)=-3.817, *MD*=-1.050, *p*=0.000

*** α=0.05 is the limit of significant level, t(df) is the t value for (df) degree of freedom,** MD, M and SD are the Mean Difference, the Mean and the Standard Deviation between the two groups respectively, and p is the significant level

General Questions

There was no significant difference between the responses of the two groups of students regarding the general questions. Specifically, both the two groups stated that they were familiar with e-learning courses and they or their teachers had used animations or simulations in their courses.

AN1 animation

The first and the second year chemistry students pointed out that they did not find the AN1 animation to be interesting and effective as a learning material. However, they mentioned that this animation could be integrated into face-to-face lecture and pre-laboratory sessions as a presentation or as an introduction to the technique by the teacher.

AN2 animation

The first year students found the AN2 animation a motivating, interesting and effective material for learning, while the second year students stated that they did not benefit much from it. Only the first year students mentioned that this animation (AN2) matched their learning style. Additionally, there was no significant difference between the first year students and the second year students' **view** regarding the integrations of the AN2 animation into face-to-face, e-learning, pre-laboratory and post-laboratory sessions. Between these study modes students preferred the integration of the AN2 animation into e-learning courses and pre-laboratory sessions.

Table 2. t-test statistical* analysis for the AN1, AN2, SIM1 and SIM2 educational materials in order to compare the two group of students' responses on the study modes (face-to-face, e-learning, pre-laboratory, post-laboratory, basic skills, blending learning)

	Face-to-Face	E-Learning	Pre-Laboratory	Post-Laboratory	Basic Skills	Blending Learning
AN1	FYS: *M*=3.7, *SD*=0.8 SYS: *M*=3.9, *SD*=0.8 *t*(38)=-0.613, *MD*=-0.150, *p*=0.543	FYS: *M*=2.6, *SD*=0.9 SYS: *M*=2.7, *SD*=1.0 *t* (38)=-0.339, *MD*=-0.100, *p*=0.736	FYS: *M*=3.7, *SD*=0.9 SYS: *M*=3.8, *SD*=1.0 *t*(38)=-0.328, *MD*=-0.100, *p*=0.745	FYS: *M*=2.7, *SD*=1.2 SYS: *M*=2.9, *SD*=1.2 *t* (38)=-0.415, *MD*=-0.150, *p*=0.680	FYS: *M*=2.4, *SD*=0.9 SYS: *M*=2.1, *SD*=0.9 *t* (38)=1.211, *MD*=0.350, *p*=0.233	FYS: *M*=3.4, *SD*=0.7 SYS: *M*=3.6, *SD*=0.8 *t* (38)=-0.627, *MD*=-0.150, *p*=0.534
AN2	FYS: *M*=2.9, *SD*=1.1 SYS: *M*=2.5, *SD*=0.8 *t* (38)=1.301, *MD*=0.400, *p*=0.201	FYS: *M*=3.9, *SD*=0.9 SYS: *M*=3.6, *SD*=1.0 *t*(38)=0.982, *MD*=0.300, *p*=0.333	FYS: *M*=3.8, *SD*=1.1 SYS: *M*=4.0, *SD*=0.9 *t*(38)=-0.940, *MD*=-0.300, *p*=0.353	FYS: *M*=2.7, *SD*=0.9 SYS: *M*=2.5, *SD*=0.9 *t* (38)=0.524, *MD*=0.150, *p*=0.603	FYS: *M*=2.6, *SD*=0.7 SYS: *M*=2.7, *SD*=1.0 *t* (38)=-0.188, *MD*=-0.50, *p*=0.852	FYS: *M*=2.9, *SD*=0.9 SYS: *M*=2.8, *SD*=0.8 *t* (38)=0.190, *MD*=0.050, *p*=0.850
SIM1	FYS: *M*=3.7, *SD*=0.9 SYS: *M*=3.7, *SD*=0.9 *t* (38)=-0.172, *MD*=-0.050, *p*=0.864	FYS: *M*=3.6, *SD*=1.0 SYS: *M*=3.9, *SD*=0.9 *t* (38)=-0.982, *MD*=-0.300, *p*=0.333	FYS: *M*=3.9, *SD*=0.9 SYS: *M*=4.0, *SD*=0.9 *t* (38)=-0.529, *MD*=-0.150, *p*=0.600	FYS: *M*=3.9, *SD*=0.8 SYS: *M*=3.8, *SD*=0.9 *t* (38)=0.551, *MD*=0.150, *p*=0.585	FYS: *M*=3.7, *SD*=0.7 SYS: *M*=3.8, *SD*=0.5 *t*(38)=-0.497, *MD*=-0.100, *p*=0.622	FYS: *M*=3.7, *SD*=0.6 SYS: *M*=3.9, *SD*=0.7 *t*(38)=-0.975, *MD*=-0.200, *p*=0.336
SIM2	FYS: *M*=4.0, *SD*=1.0 SYS: *M*=4.2, *SD*=0.8 *t*(38)=-0.677, *MD*=-0.200, *p*=0.503	FYS: *M*=2.6, *SD*=0.8 SYS: *M*=4.1, *SD*=0.6 *t*(38)=-6.869, *MD*=-1.550, *p*=0.000	FYS: *M*=2.8, *SD*=1.0 SYS: *M*=4.2, *SD*=0.7 *t*(38)=-5.445, *MD*=-1.450, *p*=0.000	FYS: *M*=3.7, *SD*=0.9 SYS: *M*=3.9, *SD*=0.8 *t* (38)=-0.796, *MD*=-0.200, *p*=0.431	FYS: *M*=3.8, *SD*=0.9 SYS: *M*=4.1, *SD*=0.7 *t* (38)=-1.125, *MD*=-0.300, *p*=0.267	FYS: *M*=4.0, *SD*=0.8 SYS: *M*=4.1, *SD*=0.7 *t*(38)=-0.420, *MD*=-100, *p*=0.6777

*** α=0.05 is the limit of significant level,** t(df) is the t value for (df) degree of freedom, MD and SD are the Mean Difference and the Standard Deviation between the two groups respectively, and p is the significant level

SIM1 simulation

Statistical analysis (t-test) showed that there was no significant difference between the two groups of students regarding their views on the interest and motivation of the SIM1 simulation for learning. The second year students found that the SIM2 simulation was more suitable for their learning style and more effective for learning. Additionally, both groups of students mentioned that the SIM1 simulation was a flexible resource which could support face-to-face, e-learning, pre-laboratory and post-laboratory session. Both groups of students believed that by using the SIM1 simulation they could obtain skills, such as observation and interpretation of experimental data, and this simulation could support blending learning courses.

SIM2 simulation

Finally, regarding the SIM2 simulation there was a significant difference between the two groups of students regarding their views on effectiveness and matching with students' learning style. Additionally, the first year found that the SIM2 simulation could be better integrated into face-to-face courses, while the second year students additionally preferred the SIM2 simulation being integrated into e-learning lectures and the pre-laboratory sessions. Both of them preferred the SIM2 simulation to be used in blended learning courses.

Correlations

For the first year students, the two computer-based applications which had a general acceptance as educational tool were the AN2 and the SIM1. Thus, correlating the six variables for the AN2 animation, we found that the variable "Interest" had a strong correlation with "Motivation", "Effectiveness" and "Students' Learning Style". On the other hand, for the SIM1 simulation the variable "Motivation" had a strong correlation with "Effectiveness" and "Students' Learning Style" (Table 3).

Overall, for the second year students, the SIM1 and SIM2 simulations had a general acceptance. By correlating the six variables for the SIM1 we found that variable "Effectiveness" had a strong positive association with the variable "Clear Interface" and with "Motivation". On the other hand, for the SIM2 simulation there are strong positive correlations between the variables "Motivation", "Clear Interface", "Students' Learning style" and "Effectiveness". Additionally, there are strong correlations between "Effectiveness" and "Clear Interface", between "Students' Learning Style" and "Clear Interface" and between "Students' Learning Style" and "Usefulness" (Table 4).

Table 3. Pearson's r analysis seeking for correlations between the six variables (usefulness, clear interface, interesting, motivation, effectiveness and students' learning style) for the first year students and the AN2 and the SIM1 learning material.

AN2	Interest-Motivation	r=0.722, p=0.000
	Interest-Effectiveness	r=0.732, p=0.000
	Interest-Students' Learning Style	r=0.671, p=0.001
SIM1	Motivation-Effectiveness	r=0.740, p=0.000
	Motivation-Students' Learning Style	r=0.585, p=0.003

r= the value of correlation and the correlation is significant at the p=0.01 level.

Table 4. Pearson's r analysis seeking for correlations between the six variables (usefulness, clear interface, interesting, motivation, effectiveness and students' learning style) for the second year students and the SIM1 and the SIM2 learning material.

SIM1	Effectiveness-Clear Interface	r=0.836, p=0.000
	Effectiveness-Motivation	r=0.845, p=0.000
SIM2	Motivation-Clear Interface	r=0.893, p=0.000
	Motivation-Students' Learning Style	r=0.677, p=0.001
	Motivation-Effectiveness	r=0.747, p=0.000
	Effectiveness-Clear Interface	r=0.784, p=0.000
	Students' Learning Style-Clear Interface	r=0.706, p=0.000
	Students' Learning Style-Usefulness	r=0.709, p=0.000

r= the value of correlation and the correlation is significant at the p=0.01 level.

Outcomes

The simplest AN1 animation required students to process only the visual information and to integrate it into their already existing knowledge schema for the titration technique. On the other hand, the most complicated SIM2 simulation demanded students to consolidate their knowledge on the topic and to perform more complicated tasks. However, the first year students preferred the AN2 animation more than the SIM1 simulation, while the second year students preferred the SIM1 more than the SIM2 simulations.

The first and the second year chemistry students had different needs, expectations and background knowledge. Usually, first year chemistry students come to University with enthusiasm which many times changes to frustration as they can not follow the overloaded curriculum or they have different expectations for their courses. The statistical analysis showed that the first year students showed preference for a computer-based application which they found to be interesting, motivating and effective for learning. The learners' characteristics and their knowledge background level played a determinative role at the use of computer-based application in chemical education. Second year students who had higher knowledge on the cognitive topic seemed to be more motivated for learning if they had a more advanced simulation than a simple application which supported low knowledge task.

The SIM2 simulation was more flexible for learning, as students could perform different experiments under different conditions as in the real laboratory and interact with their previous knowledge and new information in order to make hypothesis and make decisions. The activities such as making hypotheses and interpreting results facilitated the activation phase of the learning process, as students participated actively in the learning process and they transferred their previous knowledge to short-term memory and actively processed and combined it with the new incoming information.

Overall the SIM1 simulation was classified by the first and the second year students as an interesting and a flexible educational tool which could support blended learning courses. Additionally, they believed that they could obtain basic skills by using this simulation. The students' views about this simulation encouraged teachers to use the material that was characterized by interactivity in their teaching approach. Even if the SIM1 was considered advanced for first year students, it boosted their interest and motivation.

The AN2 animation seemed to be a more acceptable learning material for the first year students, as it visualized the technique and explained each step of it. The first year students who did not have so much experience of the technique as the second year students preferred a step-by-step animation accompanied with an explanatory text. The second year students preferred to perform virtual experiments rather than to attend and process only the visual information which was related to experimental procedure.

Finally, the AN1 animation seemed not to satisfy either the first or the second year students. Both of the groups believed that this animation could be used in face-to-face lectures where teachers explained the technique.

Overall, it was obvious that computer-based applications had different learning outcomes depending on the learners' characteristics and their background knowledge on the topic. Students with higher knowledge on a cognitive topic seemed not to be particularly assisted by a computer-based application supported by a low knowledge task.

ONLINE PRE-LABORATORY TRAINING COURSE ON VLE

In this section, an online teaching approach, which supports pre-laboratory training courses and is based on a Virtual Learning Environment, is presented. It is suitable for teachers who have classes with large number of students and limited lecture time. The theoretical background, which supports the online teaching approach, is based on constructivism, cognitive load theory and computer-supported collaborative learning.

Aim and Objectives

The aim of this teaching approach is to describe how animations, simulation, online formative computer-marked assessment and communication tools can enhance the first and second year general pre-laboratory training course designed on a Virtual Learning Environment.

Specifically, the objectives are:

- to prepare first and second year chemistry students for their practical work by filling the gap between general chemical concepts and mathematical skills and
- to enhance the formative assessment by providing feedback to the students and thus promote independent learning.

Description

Although the first year students enter an English University upon their successful A-level examination (mainly British students) or pre-university's examination (overseas students), university teachers claim that the first year students often have difficulties in topics they should well be familiar with, such as chemical concentrations and units (Royal Chemistry Society, 2007). The second year students, on the other hand, have already one year of laboratory experience, and they are called to perform more complicated experiments and interpret their experimental data during or after the laboratory session. The difficulties that the second year chemistry students usually face arise from the fact that they have not consolidated fundamental chemical concepts and techniques during their first year, such as acid-base titration, before they perform more complicated experiments such as titrations using a pH-meter, where knowledge about the structure of the glass electrode, calibration and buffer solutions is required. Additionally, in their second year practical work, students should process and analyse experimental data and interpret the values to draw out chemical conclusions during the laboratory or after that as coursework.

In order to better prepare first and second year students for their laboratory work, the teachers designed online general pre-laboratory training courses in which the following topics were covered: Safety Instructions, Glassware, SI Units, Preparing Solutions, Redox Titration and Thin Layer Chromatography (for first year chemistry students) and Experimental Error, Statistics, Calibration and Analytical Methods (for second year chemistry students). "Safety Instructions" and "Glassware" assisted the first year chemistry students to familiarise themselves with the laboratory environment, "Si Units" and "Preparing Solutions" assisted them to make unit conversions and to calculate the chemical concentrations. Finally, "Redox Titration" and "Thin Layer Chromatography" helped the first year chemistry students to perform virtual experiments by using simulators and to make the appropriate calculations based on these specific techniques. The second year pre-laboratory training course was focused more on the experimental data analysis (experimental error and statistical analysis) and the techniques (calibration and analytical methods such as spectroscopy and electrochemistry).

The online theory, which was included in these pre-lab courses, was limited, and only the most important points were emphasised. For example, the text included "Notes" and "Useful Tips", so that students were aware in advance of the most important theoretical parts of the experiments, before they performed them in the laboratory. Images with clear messages but without using a lot of colours were integrated into the text. The topics included, apart from the text and static images, animations, virtual simulations, self-assessment exercises supported by feedback and final computer-marked formative assessments (Figure 4). Thus, the students could read the learning material, which was uploaded by the teacher, perform virtual experiments, participate in the self-assessment by instantly taking feedback for each question according to their response and then finally completed the final computer-marked formative assessments. Finally, by using communication tools such as e-mail, students could contact the teacher or one of their peers seeking assistance on specific misunderstandings or difficulties that they faced.

Additionally, in the first year chemistry pre-laboratory training course, animations supported by text and simple simulations were used, while in the case of the second year pre-laboratory courses, simulations and suitable tasks were available for students' courses. The choice of the multimedia learning formats that supported the online training courses was based on the outcome of previous investigations, as presented in the section "Students' perspectives for computer-based applications" of this chapter.

Figure 4. Part of the inside appearance of each topic which includes theory, virtual experiments, self-assessment with feedback and the final assessment.

Students were free to participate in the on-line training any time they wanted within a period of one or two week(s) depending on the topic's difficulties. The students could submit their responses on the assessments electronically one day before the laboratory session begun. In the laboratory, the students performed the experiments without the additional instructions from the teacher.

Outcomes

The aim of the pre-laboratory course was not to stress the already overloaded curriculum, but to enhance the teaching approach by giving students the opportunity to follow any learning avenue they saw fit, depending on their learning style. For that purpose in the pre-laboratory training courses apart from theory parts, multimedia resources (animation and simulation), self-assessment and formative assessment provided the students' preparation. The background theory was presented in a variety of ways, which made the learning environment flexible for students to follow. For example, by using conceptual animations students were able to make connections among the macroscopic, microscopic and symbolic level of representations. The self-paced, online pre-laboratory course assisted students to review concepts, fill the gaps between the theory and practical work, build confidence before entering the laboratory and strengthen their mathematical skills. The virtual simulators gave them the opportunity to process and store information about the experimental configuration, the basic concept and the calculations and to make the connections between their prior knowledge before entering the laboratory.

In these online courses, teachers did not have the same level of control on teaching as with the traditional lecture approach. The students used animations, simulations, pictures and diagrams in order to facilitate their own learning process and formative assessment supported by feedback allowing them to take control of their own learning. Adopting this teaching approach proved to be quite a difficult challenge for teachers, as they had to carefully consider the students' background and their experience of practical work in order to create generally acceptable learning material for students and give them the flexibility to adopt the learning material to their learning style.

By providing feedback to students, the teachers could enhance students' learning and promote independent learning, as students could by themselves identify errors and practice the skills that they needed. Feedback promoted self-directed learning and students were encouraged to solve their tasks by applying critical thinking rather than memorising the solution.

Communication tools were very helpful in facilitating the teacher's supervision of learning. For example, if the teacher wanted to send a general instruction, (s)he could do so easily online and thus maintain a constant communication with the students. Using this technology the teacher prevented students from being frustrated when they could not solve a problem or they had difficulties understanding a topic, by giving them some online tips. Additionally, through communication technology the teacher could discover more easily the students' background and make appropriate changes to the learning material to better adapt it to the students level. Learning is a difficult, active and continuing process and students should not feel frustrated at any point by their course as this could result in them giving up. A specific example based on the collaboration between students and teachers designed on Virtual Learning Environment in combination with students' evaluation is presented in the journal Computers & Education (Limniou et al., 2009).

Examples for Face-to-Face Activities

In this section, two methodological procedures for the support of face-to-face blended learning activities in science courses are proposed using virtual simulation as an educational tool. The theoretical background which supports the proposed methodologies is based on constructivism, cognitive load theory and computer-supported collaborative learning.

Example 1

Aim and Objectives

A scientific instrument converts non-electrical physical and chemical characteristics such as temperature, intensity of light, chemical composition and pressure into electrical signals such as current, voltage, or charge that may be manipulated and interpreted by a human. A sensor is a device that receives and responds to a signal or a stimulus; therefore a sensor detects information and/or controls a system. A sensor that is often used in scientific instruments is the thermistor. Changes in temperature cause changes in its resistance. Many researchers have described the design and construction of a computer-controlled instrument in order to decrease the cost of laboratory activity, to collect results with more accuracy, to make experiments less hazardous for the students, make students familiar with modern instrumentation and allow students to focus on results rather than tedious measurements (Papadopoulos & Limniou, 2002; Steehler, 1998). Before the students use the thermistor in their experiments, it is important that they understand its operation and how they can calibrate it (Foster, 2005; Wong et al., 2001; Muyskens, 1997). The aim of this methodology is to suggest an instrumentation course based on a computer-assisted teaching approach where the calibration concept is combined with a flexible virtual learning environment. As an educational tool, a virtual thermistor simulator is used. The objectives of this methodological procedure for students are:

- to feel more confident with modern laboratory instrumentation,
- to be familiar with calibration theory and put it into practice on the virtual thermistor simulator,
- to perform experiments and receive data as part of the process of learning how to use a thermistor in the laboratory,
- to obtain computer experience by giving them the feel of working on modern instrumentation and
- to obtain skills such as processing and analysis of experimental data

Description

The instrumentation course was conducted in the laboratory which was equipped with Personal Computers (PCs) and the students were divided into two-person teams who worked on a single PC. At the beginning of the lecture, the teacher explained the calibration concept to students by posing questions to them. By processing the student's answers, the teacher had the opportunity to identify misconceptions that the students might have had. Additionally, the teacher described the calibration concept and procedure by using terms, graphs, and mathematical equations in order to encourage group discussion and exchange of ideas.

In the next stage of the teaching approach, the teacher introduced the Thermistor simulator to students in order to familiarise them with the simulation environment and its basic functionalities. As the teacher performed the virtual experiments and at the same time explained his/her steps, (s)he enabled students with a visual style of learning to receive information through visualisation, while students with verbal style of learning received information through his/her explanations. The purpose of these experiments was not only for students to familiarise themselves with a simulator before use it but to better understand the calibration concept.

The NTC Thermistor Simulator, which is the educational tool under discussion, is a Windows compatible computer program that simulates a simple use of a thermistor in a laboratory. The main part of the simulator illustrates a thermometer, a beaker, a heat source, magnifying lens and a box in which there are the thermistor displays (voltage, resistance and temperature). Students had to virtually turn on the power of heat source to increase the temperature and the red line inside the thermometer was lengthening as in a real thermometer. The display inside the magnifying lens changed and recorded the temperature of the thermometer. Students could choose one of the three NTC Thermistors or add a new thermistor (Figure 5). The simulation program has been designed to give to both students and teacher the opportunity to add a new thermistor and its characteristics. Additionally, the simulator can be interfaced with an instrument through the option "Add a new Thermistor" as well.

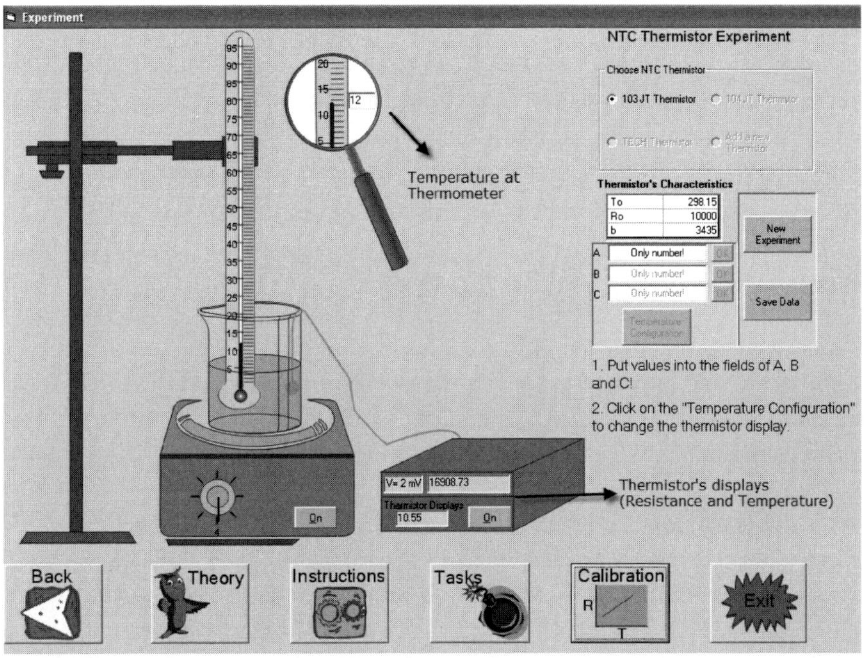

Figure 5. Part of the simulation which illustrates the basic characteristics of the virtual simulation.

After choosing a thermistor, students had to observe the changes on the thermistor's temperature display and on the resistance display, as the temperature was increasing. Students could witness the difference of the two temperature values (thermometer and thermistor displays), so in essence they experienced first hand the notion of calibrating the thermistor. Students could save the experimental data at any temperature, and later process them on the spreadsheet such as Microsoft Excel. Following the Steinhart-Hart equation (1), students then calibrated the thermistor using the equation:

$$1/T = A + B\ln R + C(\ln R)^3 \qquad (1)$$

where: T is temperature in Kelvin, R is the thermistor resistance and A, B and C is the curve-fitting constants

The students worked in two-person teams had to make all the appropriate calculations in order to calibrate a specific thermistor for a certain temperature range. To achieve this goal students had to collaborate with their team member.

At the next step of the teaching approach the teacher used the simulator to interface it with a real thermistor under the option “Add a new Thermistor”. A real thermistor computer-controlled system was presented, and students had the opportunity to pose questions to the teacher on this newly presented system, perform real experiments and collect real experimental data.

When all the students finished setting up and calibrated their thermistor (virtual and in a computer-controlled system), the teacher asked them to draw and explain how a thermistor was interfaced with the computer. The aim of this part of the lecture was for students to think how the scientific instrument was interfaced with computers and to summarise the calibration of the thermistor.

At the final stage of the lecture the teacher tried to engage students in a discussion on how different sensors can be used in science and how they are interfaced with a computer. Communication between students and teacher was at the centre of collaboration, be it to exchanging points view, debate disagreements or explain difficult concepts.

Outcomes

The role of the teacher in the above course was not to provide specific answers to the students but to provide alternate avenues for students to understand the topic depending on their own learning styles. By using the simulator, the students had the opportunity to practice their understanding in the real instruments and the lecture could be expanded from a specific example (thermistor calibration) to general examples (sensor calibration). The key point is that students did not follow the teacher’s instructions but it was the teacher who followed the students’ learning style through face-to-face discussion, virtual and real experiments.

The learning environment was designed to be flexible and motivate students to think of modern instruments and their calibration in a way, which otherwise would not have been feasible in the available laboratory time. Following the above procedure both students with verbal and visual style of learning responded positively through visualisation by using simulation and verbal interactions via collaboration (student-student and student-teacher). All students started to work initially as two-person teams, then as a four-person team and finally all students together as a whole group. The exercises were purposely selected to become sequentially more difficult, starting from small and specific tasks and continued to bigger and more complicated tasks. Additionally, students were able to select a series of icons such as theory, task, instruction and calibration in order to read the supplementary material. The teacher could change icons, part of the theory and tasks and add the thermistor that (s)he wanted to adjust the visual simulation to the needs of his/her lecture.

Also, by using the simulation program, students could save their data as an ASCII data file and import into spreadsheets such as Microsoft Excel, therefore, they could familiarized themselves with processing data and build on their IT skills. More detailed examples based on the above methodology are presented in the journal Educational and Information Technologies (Limniou et al., 2007) and the journal Chemistry Education Research and Practice (Limniou et al., 2007).

EXAMPLE 2

Aim and Objectives

In the science curriculum, there are often difficult topics that teachers cannot extensively cover due to limited lecture or laboratory time. For example, the curriculum devotes a substantial amount of lecture time to some important cognitive topics which students practice in various laboratory projects. Students should develop specific skills during their studies; such as interpreting experimental data and graphs. In physics education, for example, students have difficulty in understanding line graphs (Ploetzner et al., 2009; Scanlon, 1998; Beichner, 1994; Berg & Smith, 1994). Traditionally, teachers cover these topics by providing different examples to students and/or allowing students to work on them. However, the limited number of examples and the lack of lecture time do not allow students to develop the specific skills which are required. By using computer-based applications, the teaching approach can be enhanced by simply providing more examples to the students and/or allowing collaboration activities between them and their teachers.

The aim of this methodology is to show how a flexible computer-based environment can be integrated into lectures, pre-laboratory or post-laboratory sessions in order for students to develop useful scientific skills. The objectives of this methodological procedure are to allow:

- students to work on different examples at their own time,
- students to receive information gradually and to start to build the skills they need from a significant number of given examples,
- the asynchronous collaboration between students and teacher and
- teachers to adjust their teaching approach according to their needs and according to the students' learning style and expectations.

Infrared (IR) spectroscopy is used as an example. Undergraduate and graduate students use infrared (IR) spectrometers and are taught how to collect information from the resulting spectroscopic data during the laboratory. This data is used to gather information about a compound's structure and the functional groups it contains, to assess its purity and to identify it. IR spectroscopy is covered through lectures and students gather the data through laboratory sessions. Many researchers have combined computational chemistry and experimental spectroscopy to create interactive displays of IR spectra for the use on the World Wide Web or they have designed teaching procedures to assist students to better understand IR spectroscopy based on computer applications using spreadsheets (Wallace & Carpenter, 2005; Buser & McDougal, 2004; Fisher & Holme, 2000; Lahti et al., 2000). Our contribution to the difficult topic of IR spectroscopy is the design and development of a simulation program based on the cognitive load theory to assist second year chemistry students to interpret the IR spectra using an interactive way of learning and also provides to the teachers a flexible method of teaching.

Description

The educational tool is an IR spectrum simulator, in which there are three categories of tasks, each including different samples. The aim of the first task is to highlight the sample's structure with the corresponding part on the graph. The relevant information on the samples structure in this theory overview is purposefully revealed in small controlled steps so that students can receive it gradually and absorb its content. In the next task, the simulation gives students the graph accompanied with basic values of peaks and students can match the peak value with the corresponding segment of the sample. Upon finishing the exercise, the simulation reveals the correct answer to students. At the third task, students interpret the data for unknown samples and send their answers to their teacher and/or their peers (Figure 6).

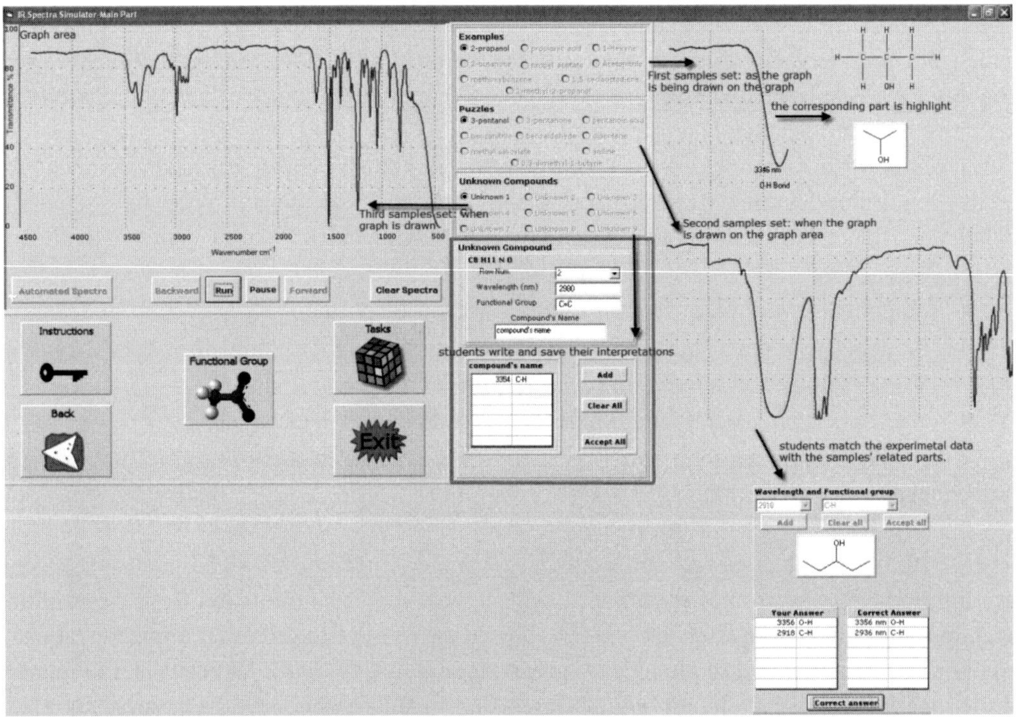

Figure 6. Part of the simulation which illustrates the basic characteristics of the virtual simulation. All the graphs are drawn in the "Graph area" and the highlights for the first samples set; matching the wavelength with functional group panel in the simulation is located at the purple place.

All the samples that were used in the IR spectra simulator could be changed by the teacher creating a flexible learning environment for students and providing different examples for each student or group of students. Specifically, by using this educational tool teachers in their lectures, pre-laboratory or post-laboratory sessions, could introduce students to IR spectra interpretation by visualising the contribution of different parts of the sample on the graph and/or to compare graphs corresponding to samples with similar characteristics.

Depending on the study mode (lecture, pre-laboratory or post-laboratory session) teachers gave either an extensive or a short presentation on IR spectra interpretation to students by using the IR simulator. During either lecture, teachers presented different examples and encouraged students to predict which part of the sample was responsible for specific peaks or to explain why that part of a sample was responsible for the matching peak. Students were divided into groups and a series of unknown samples were assigned to them for interpretation.

In the case of the pre-laboratory session, teachers followed a different teaching approach. Before the pre-laboratory session, teachers assigned to each group of students different samples and they encouraged them to work as a team in order to interpret the spectra and identify the unknown samples. An online discussion took place through e-mails. During the pre-laboratory session students presented their findings to the whole class and a face-to-face discussion took place.

In post-laboratory session, students used the experimental data which they have gathered during the laboratory and they discussed together the interpretation after uploading to IR spectra simulator.

In all the above study modes (lecture, pre- and post-laboratory session), teachers assigned to different group of students different examples. Students could send their responses or pose questions to teachers or their students through e-mails. Students had to send their final responses to teachers one day before their next lecture or laboratory. Teachers facilitated the students' questions or comments twice a week. The first facilitation day was three days after the lecture or laboratory and the second was the day before the next week's lecture or laboratory.

A day before the lecture or laboratory teachers had already collected all the students' responses on the tasks and they could summarise the cognitive topic and clarify students' misconceptions which were related to their responses.

Outcomes

One of the necessary skills that students should obtain during their studies is the interpretation of spectra. Usually, during a lecture or a laboratory session students could not practice interpretation, as they did not have time to observe many graphs from different samples. By using a flexible simulator where all the samples could be changed by the teacher, students had the opportunity to study many graphs and to observe which peaks were significant and where they should focus on interpreting an unknown spectra. The three different kinds of tasks assisted them to understand step-by-step the interpretation of the IR spectrum.

During the introduction part, teachers presented varying levels of tasks and provided enough time to students to store the new information in their long term memory. The first step was for students to make the visual connections between their previous knowledge about the samples and the new information about their characteristic peaks in the graph. Teachers and students discussed together using the simulator's samples. When teachers assigned to group of students different tasks on the same simulator, teachers gave students the opportunity to

- think about graphs (peaks and corresponding samples' parts) at their own pace,
- pose questions to other members of their teams or to all their peers and teachers,
- an asynchronous discussion in order to exchange views with their peers,

- collaborate as a team in order to submit their final responses and
- participate actively in learning.

The simulator and the way that was integrated into lectures, pre-laboratory or post-laboratory sessions had special merits on feedback, interactivity, and flexibility. Students got feedback from the simulator, their teachers and their peers through e-mails. Teachers set up different tasks regarding the students' background or their lecture needs.

Teachers had a vision of the learning approach, where students learned from each other and from examples instead of being told by the teacher. All participants in discussions (teacher and students) had a constructive dialogue by giving feedback to each other and the simulator was the medium to illustrate different examples. This procedure was different from the conventional way of teaching where feedback was provided only by the teacher. During discussion, students constructed their new knowledge through addition, explanation, evaluation and summary.

According to Veerman et al. (2000), during the addition process an input of new information is linked to the discussion. During the explanation process, information is differentiated, specified, categorised, or made clear by examples. Evaluations are (personally) justified considerations of the strength or relevance of already added or explained information. During summary the already given information is reorganised or restated in such a way that the main points or (sub)conclusions reflect the discussion. More examples for computer-supported collaborative learning and teachers' role are presented in the book chapter "The role of Simulations and Real-time applications in Collaborative Learning" (Limniou et al. 2009).

Discussion

Computer-assisted teaching is becoming a popular approach in Higher Education around the world and many researchers have studied how Information and Communication Technology (ICT) can enhance teaching and learning. The aim of this chapter is to study the application of the computer-based teaching on Science Education by using examples from Chemistry Education. In order to effectively design a computer-based application for learning, teachers should be aware of cognitive load theory. Other parameters such as students' characteristics (background knowledge level, learning style etc.) and the integration of the applications into courses should be studied by teachers, before they follow a computer-assisted teaching methodology. In this chapter, methodologies which are concerned on online training courses or face-to-face lectures are described. The key points in all proposed methodologies were the following:

1. Learning environment follows cognitive load theory

The structure and the interface of computer-based applications facilitated students to find information easily and quickly. Icons reduced the complexity of the system, making it easier to learn and use. The virtual objects on the screen had characteristics similar to the real objects, so that learners needed little time to perceive and to recognise them. The computer-based applications provided the learners with information gradually without overloading their working memory.

2. Flexible learning environments

The learning environment was characterised by flexibility. Icons, tasks and theory parts, which supported the applications, could be changed by teachers to create educational tools suitable for their teaching. Thus, teachers could easily customize those computer-based applications which met the specific needs of individual learners, adapted to students' background and/or covered their teaching needs.

3. Collaborative learning

Collaborative learning supports the use of the effective learning methods (e.g. make explicit, discuss, reason), while allowing for the acquisition of essential social communications skills (Kirschner, 2001). Through collaboration students could clarify misconceptions or be taught by peer interaction, as they exchanged views on a specific topic or task. Synchronous and asynchronous collaborations presented different characteristics not only on the views of place and time, but on the perspectives of the students' and teachers' roles. For example, in the presented synchronously methodological procedures, students had the opportunity to spontaneously ask questions, react to what happens and to perceive how other members react. The asynchronous collaboration gave students the opportunity to have more time to prepare their comments. Additionally, students who did not like to express themselves orally during lectures could participate actively in the asynchronous collaboration.

4. Feedback

Students participated in the teaching and learning procedure actively by receiving feedback from interactions with their peers or the teacher or the computer-based applications. Additionally, authentic tasks or assessments from science could be used to link learning and work by creating a correspondence between what was assessed in the university with what students needed to do in their employment after finishing their education. *These authentic assessments can a) stimulate students to learn deeply b) stimulate students to develop professionally relevant skills and thinking processes used by professionals and c) motivate students to learn by showing the immediate relevance of that what is learnt for professional practice* (Gulikers et al., 2008). In all the proposed methodologies students had the opportunity to work on authentic tasks and the feedback that they received was based on their interaction with their teacher, their peers and the learning material (animations, simulations, formative assessment supported by feedback etc.).

5. Teacher

Learning is aided by conversation that seeks and clarifies the ideas of learners (Shiland, 1999). In the proposed methodologies, the role of the teacher was to facilitate by encouraging students to participate actively in the teaching procedure through synchronous and/or asynchronous discussion. Students could exchange views and experimental data and one could teach the others. The teacher asked them to generate questions, predictions and explanations. Questions were posed by the teacher in order to create dissatisfaction with the learners' present knowledge. The common misunderstanding that usually teachers had regarding the online courses was that their involvement finishes when uploading the learning material. However, teachers should participate actively in online discussions by keeping the conversion live. If students' questions remained unanswered, then they felt disappointed.

6. Process information

Learning is an active process, and learners necessarily process information actively to comprehend it (Ausubel, 1960). Following the above methodologies, students received the information gradually having enough time to process it and to integrate it into the appropriate cognitive schema. The teacher posed questions to the students or assigned tasks to them in order to create dissatisfaction with their present knowledge or encouraged them to identify variables, state problems, make hypothesis, and generate questions, predictions or explanations.

7. Motivation

Group motivation is often determined by factors such as sharing the same goals (Slavin, 1994), and *personal motivation is determined by factors such as active participation and personal responsibility* (Salomon & Perkins, 1998). In our proposed methodologies, motivation was reflected by a high level of active participation of the students in learning activities and an individual willingness to be taught by personal curiosity and/or interest in a specific topic. Aspects such as students' control of learning, immediate feedback, and opportunity to generate questions, increased their motivation.

8. Learning style

In the proposed methodologies the learning environments provided interactivity and rapid access to feedback for the students and displayed multiple representations thus enhancing active learning through exploration. The information which was presented in a variety of ways such as images, animation, simulation and activities such as face-to-face interactions with the teacher, collaboration with their peers and interaction with simulations allowed students to receive the same information from different learning "avenues". By using a plethora of different learning venues, students with different learning styles could equally absorb information.

The combination of online and face-to-face instruction methods is known as blended learning and it is differentiated from other formats based upon the percentage of time spent online. The proposed methodologies described different integrations of computer-based applications into chemistry education presenting different activities from face-to-face to asynchronous. In all of the above methodologies, there were benefits for enhancing of educational procedure, as teachers and students shared innovations and experience making the learning strategies more economically feasible than the conventional teaching. Additionally, students did not have dissatisfactions, such as absence of a learning atmosphere and lack of interpersonal and direct interactions among students and teachers, which they describe for e-learning courses (Liaw, 2008). Thus, teachers could create new teaching approaches based on blended learning depending on minimum required space and time, financial difficulties, interactions with students enriching it with different delivery of new information, where students can follow their learning style and strategy. Blended learning effectiveness can be influenced by multimedia instruction, interactive learning activities and learning environment.

Apart from the traditional teaching on science education, there are other different approaches based on inquiry and discovery. Although the main advantage of traditional teaching is that can accommodate the large number of university students in limited lecture and laboratory time, following any other teaching approaches, students can be more familiar with the scientific way of thinking. In computer-assisted teaching, students have the benefit to participate actively in learning and pore over the scientific way of thinking, as they make observations and hypotheses, process, analyse and interpret experimental data. For example, teachers encourage students to generate hypotheses and perform the virtual experiments in order to validate these by using simulations. Inquiry-based learning can lead to a deeper and more meaningful understanding, but its effectiveness in learning depending on learners' prior knowledge. Thus, through computer-assisted teaching teachers provide online material to students or discuss with them scientific concepts by using computer-based applications having as the aim to clarify their misconceptions preparing themselves to accept the new knowledge. Students generate their own interpretations based on their background, attitudes, abilities, experiences etc, before, during and after instruction. The result of the above procedure is that students learn the scientific way of thinking and teachers create a flexible learning environment, where the large number of university students and the limited lecture and laboratory time are not obstacles in teaching. Teachers can adopt the basic concepts of any teaching approaches and adapt them to computer-assisted teaching partially or totally by using computer-based applications and creating appropriate activities for students.

CONCLUSION

Teaching and learning at university level of science is challenging due to the complex and abstract nature of science concepts and the experience on laboratory. Problem-based, discovery and inquiry activities focused on real world experience and the use of the visualisation tools and the virtual simulations enhance the understanding of scientific ideas and phenomena among students. In computer-assisted teaching apart from computer-based applications teachers play the most important role. Teachers need to have imagination in order to integrate computer-based applications into their teaching approach by resolving difficulties in scientific concepts to students, promoting collaboration between students and assisting them to construct their own knowledge. Teachers' strong background on pedagogy, psychology and science are necessary for a successful integration of computer-based applications into teaching methodologies. As usually teachers are mainly qualified in their scientific field, they need training courses in workshops or in their workplaces (Galanouli et al., 2004) or a joint effort of instructors, education-researchers and computer engineers in design web-based courses or multimedia formats is necessary (Barak, 2007). In computer-assisted teaching, the learning procedure can be conducted in a flexible and pleasant environment, in which students can meet their needs. According to the proposed methodologies, learners generated their own understanding based on their background and their learning style. Since students built their own concepts, their constructions differ from the one that the instructor held and had tried to present. Technology applications could enhance the learning process and outcome by facilitating the implementation of the instructional strategy and by eliciting and engaging the underlying cognitive process of learning.

REFERENCES

Alavi, M. & Leidner, D. E. (2003). Virtual Learning Systems. *Encyclopaedia of Information Systems, 4*, 561–572.

Ausubel, D. P. (1960). The use of advance organizers in the learning and retention of meaningful verbal materials. *Journal of Educational Psychology, 51(5),* 267–272.

Ausubel, D. P. (2000). *The acquisition and retention of knowledge: a cognitive view.* Boston: Kluwer Academic Publishers.

Bailey, P. D. & Garratt, J. (2002). Chemical Education: theory and practice, *University Chemistry Education, 6(2),* 39–57.

Bannert, M. (2002). Managing cognitive load-recent trends in cognitive load theory. *Learning and Instructions, 12(1),* 139–146.

Barak, M. (2007). Transition from traditional to ICT-enhanced learning environments in undergraduate chemistry course. *Computers & Education, 48(1),* 30–43.

Beichner, R. J. (1994). Testing student interpretation of kinematics graphs. *American Journal of Physics, 62(8),* 750–752.

Berg, C. A. & Smith, P. (1994). Assessing students' abilities to construct and interpret line graphs: Disparities between multiple-choice and free-response instruments. *Science Education, 78(6),* 527–554.

Betrancourt, M., Dillenbourg, P. & Clavien, L. (2008). Display of Key Pictures from Animation: Effects on Learning. In Rouet, J. F., Lowe, R. & Schnotz, W. (Eds.), *Understanding Multimedia Documents* (pp. 61–76). New York: Springer Science-Business Media.

Blas, T. M. & Serrano-Fernández, A. (2009). The role of new technologies in the learning process: Moodle as a teaching tool in Physics. *Computers & Education, 55(1),* 35–44.

Bliuc, A. M., Goodyear, P. & Ellis, R. (2007). Research focus and methodological choices in studies into students' experiences of blended learning in higher education. *The Internet and Higher Education, 10(4),* 231–244.

Bodner, G. M. (1986). Constructivism: A theory of knowledge. *Journal of Chemical Education, 63(10),* 873–878.

Bodner, G., Klobuchar, M., & Geelan, D. (2001). The many forms of constructivism. *Journal of Chemical Education, 78(8),* 1107–1115.

Boucheix, J. M. & Schneider, E. (2009). Static and animated presentations in learning dynamic mechanical systems. *Learning and Instructions, 19(2),* 112–127.

Bunce, D. M. (2001). Does Piaget Still Have Anything to Say to Chemists?. *Journal of Chemical Education, 78(8),* 1107–1121.

Bunce, D. M., VandenPlas, J. R., & Havanki, K. L. (2006). Comparing the effectiveness on student achievement of a student response system versus online WebCT quizzes. *Journal of Chemical Education, 83(3),* 488–493.

Buser, J. Y. & McDougal, O. M. (2004). A Pedagogical Presentation of Data to Supplement the Instruction of Organic Spectroscopy. *The Chemical Educator, 9(4),* 216–219.

Chandler, P. & Sweller, J. (1991). Cognitive Load Theory and the Format of Instruction. *Cognition and Instruction, 8(4),* 293–332.

Chang, K. E., Chen, Y. L., Lin, H. Y. & Sung, Y. T. (2008). Effects of learning support in simulation-based physics learning. *Computers & Educations, 51(1),* 1486–1498.

Clow, D. (1998). Teaching, Learning and Computing. *University Chemistry Education, 2(1),* 21–28.

De Jong, T., Martin, E., Zamarro, J. M., Esquembre, F., Swaak, J. & van Joolingen, W. (1999). The integration of computer simulation and learning support: An example from the physics domain of collisions. *Journal of Research in Science Teaching, 36(5),* 597–615.

De Jong, T. & Van Joolingen, W. R. (1998). Scientific discovery learning with computer simulations of conceptual domains. *Review of Educational Research, 68(2),* 179–201.

Domin, D. S. (1999). A Review of Laboratory Instruction Styles. *Journal of Chemical Education., 76(4),* 543–547.

Felder, R. M. & Silverman L. K. (1988). Learning and Teaching Styles In Engineering Education. *Engineering Education, 78(7),* 674–681.

Felder, R. M. & Brent, R. (2005). Understanding Student Differences. *Journal of Engineering, 94(1),* 57–72.

Fisher, L. & Holme, T. (2000). Using Web-Based Databases in Large-Lecture Chemistry Courses. *The Chemical Educator, 5(5),* 269–276.

Foster, B. L. (2005). Mercury Thermometer Replacements in Chemistry Laboratories. *Journal of Chemical Education, 82(2),* 269–273.

Gabel, D. (1999). Improving Teaching and Learning through Chemistry Education Research: A Look to the Future, *Journal of Chemical Education., 76(4),* 548–554.

Galanouli, D., Murphy, C. & Gardner, J. (2004). Teachers' perceptions of the effectiveness of ICT-competence training. *Computers & Education, 43(1–2),* 63–79.

Garratt, J. (1997). Virtual investigations: Ways to accelerate experience. *University Chemistry Education., 1(1),* 19–27.

Gulikers, J. T. M., Kester, L., Kirschner, P. A. & Bastiaens, T. J. (2008). The effect of practical experience on perceptions of assessment authenticity, study approach, and learning outcomes. *Learning and Instruction, 18(2),* 172–186.

Gunawardena, C. (1995). Social presence theory and implications for interaction and collaborative learning in computer conferences. *International Journal of Educational Telecommunications, 1(2),* 147–166.

Gunstone, R. F. & Champagne, A. B. (1990). Promoting conceptual change in the laboratory. In E. Hegarty-Hazel (Ed.), *The student laboratory and the science curriculum* (pp. 159–182*)*. London: Routledge.

Homer, B. D., Plass, J. L. & Blake, L. (2008). The effects of video on cognitive load and social presence in multimedia-learning. *Computers in Human Behavior, 24(3),* 786–797.

Johnstone, A. H. (1997). Chemistry Teaching—Science or Alchemy?. *Journal of Chemical Education., 74(3),* 262–268.

Kirschner P. A. (2001). Using integrated electronic environments for collaborative teaching/leaning. *Research Dialogue in Learning and Instruction, 2(1),* 1–9.

Lahti, P. M., Motyka E. J. & Lancashire, R. J. (2000). Interactive Visualization of Infrared Spectral Data: Synergy of Computation, Visualization, and Experiment for Learning Spectroscopy. *Journal of Chemical Education, 77(5),* 649–653.

Liaw, S. S. (2008). Investigating students' perceived satisfaction, behavioural intention, and effectiveness of e-learning: A case study of the Blackboard system. *Computers & Education, 51(2),* 864–873.

Limniou, M., Papadopoulos, N., Giannakoudakis, A., Roberts, D. & Otto, O. (2007). The integration of a viscosity simulator in a chemical laboratory. *Chemistry Education Research and Practice, 8(2)*, 220–231.

Limniou, M., Papadopoulos, N. and Roberts D. (2007). An integrated lecture, virtual instrumentation lab approach to teaching UV-Vis spectroscopy. *Educational and Information Technologies, 12 (4)*, 229–244.

Limniou, M., Papadopoulos, N. & Whitehead, C. (2009). Integration of simulation into pre-laboratory chemical course: Computer cluster versus WebCT. *Computers and Education, 52(1)*, 45–52.

Limniou, M., Papadopoulos, N. & Kozaris, Y. (2009). The role of Simulations and Real-time applications in Collaborative Learning. In Luzzatto, E. & DiMarco, G. (Eds) Collaborative Learning: Methodology, Types of Interactions and Techniques, New York: Nova Science Publishers, in press.

Lowe, R. K. (2003). Animation and learning: Selective processing of information in dynamic graphics. *Learning and Instruction, 13(2)*, 157–176.

Lowe, R. K. (2004). Interrogation of a dynamic visualization during learning. *Learning and Instructions, 14(3)*, 257–274.

Mayer, R. E., & Moreno, R. (2000). A Learner-Centered Approach to Multimedia Explanations: Deriving Instructional Design Principles from Cognitive Theory. *Interactive Multimedia Electronic Journal of Computer-Enhanced Learning, 2,* 1–6.

Mayer, R. E. (2001). Multimedia-learning, New York: Cambridge University Press.

Mayer, R. E. & Moreno R. (2002). Aids to computers-based multimedia learning. *Learning and Instructional, 12(1)*, 107–119.

Mayer, R. E & Moreno, R. (2003). Nine ways to reduce cognitive load in multimedia learning. *Educational Psychologist., 38(1)*, 43–52.

Matthews, R. S., Cooper, J. L., Davidson, N. & Hawkes, P. (1995). Building bridges between cooperative and collaborative learning. *Change, 27(4)*, 34–37.

Merritt, M. V.; Schneider, M. J.; Darlington, J. A. (1993). Experimental Design in the General Chemistry Laboratory. *Journal of Chemical Education., 70(8)*, 660–662.

Muyskens, M. A. (1997). An Integrated-Circuit Temperature Sensor for Calorimetry and Differential Temperature Measurement. *Journal of Chemical Education, 74(7)*, 850-852.

National Research Council (1996). *National Science Education Standards*, National Academic Press: Washington D.C.

Ngai, E. W. T., Poon, J. K. L., & Chan, Y. H. C. (2007). Empirical examination of the adoption of WebCT using TAM. *Computers & Education, 48(2)*, 250–267.

Nicholls, B. S. (1999). Pre-laboratory support using dedicated software. *University Chemistry Education, 3(1)*, 22–27.

Nicholls, B. S. (1998). Post-laboratory support using dedicated courseware. *University Chemistry Education, 2(1)*, 10–15.

O'Connor, S. & Ross, A. (2004). WebCT role-playing: Immediacy versus e-Mediacy in learning environments. *Learning Environments Research, 7(2)*, 183–201.

Paas, F. G. W. & Van Merriënboer, J. J. G. (1994). Variability of worked examples and transfer of geometrical problem solving: A cognitive load approach. *Journal of Educational Psychology, 86(1)*, 122–133.

Palmer, S. (2007). An evaluation of streaming digital video resources in on- and off-campus engineering management education. *Computers & Education, 49(2)*, 297–308.

Papadopoulos, N. & Limniou, M. (2002). How to Use Visual Basic to Interface Scientific Instruments to a Personal Computer. *The Chemical Educator., 7(5)*, 288–291.

Papadopoulos, N. & Limniou, M. (2003). pH Titration Simulator. *Journal of Chemical Education, 80(6)*, 709.

Ploetzner, R., Lippitsch, S., Galmbacher, M., Heuer, D. & Scherrer, S. (2009). Students' difficulties in learning from dynamic visualisations and how they may be overcome. *Computers in Human Behavior, 25(1)*, 56-65.

Robinson, W. R. (2004). Cognitive Theory and the Design of Multimedia Instruction. *Journal of Chemical Education, 81(1)*, 10–13.

Rouet, J. F., Lowe, R. & Schnotz, W. (2008). *Understanding Multimedia Documents* (1st Edition). New York: Springer.

Royal Society of Chemistry (2007). School-to-university transition: comparison of the A-level chemistry specifications with year 1 at university. *Chemistry in our Future.* http://www.rsc.org/images/SyllGpFinal_tcm18-110563.pdf

Salomon, G. & Perkins, D. N. (1998). Individual and social aspects of learning. *Review of Research in Education, 23(1)*, 1–24.

Scanlon, E. (1998). How beginning students use graphs of motion. In van Someren, M. W., Reimann, P., Boshuizen, H. P. A. & de Jong, T. (Eds), *Learning with multiple representations* (pp. 67–86). Amsterdam: Pergamon Press.

Schnotz, W. (2008). Why multimedia learning is not always helpful. In Rouet, J. F, Lowe, R. and Schnotz W. (Eds), *Understanding Multimedia Documents* (pp. 22), Springer: New York.

Schnotz, W. & Kürschner, K. (2007). A Reconsideration of Cognitive Load Theory. *Educational Psychology Review, 19(4)*, 469–508.

Schnotz, W. & Rasch, T. (2005). Enabling, Facilitating, and Inhibiting Effects of Animations in Multimedia Learning: Why Reduction of Cognitive Load Can Have Negative Effects on Learning. *Educational Technology Research and Development, 53(3)*, 47–58.

Slavin, R. (1994). Cooperative learning: theory, research, & practice (2nd Edition). Boston: Allyn & Bacon.

Seng, L. & Mohamad, F. S. (2002). Online learning Is it meant for science courses?. *The Internet and Higher Education, 5(2)*, 109–118.

Shiland, T. W. (1999). Constructivism: the implication for laboratory work. *Journal of Chemical Education, 76(1)*, 107–108.

Steehler, J. K. (1998). Should Advanced Instruments Be Used in Introductory Courses?. *Journal of Chemical Education, 75(3)*, 274–275.

Sweller, J. (1994). Cognitive load theory, learning difficulty, and instructional design. *Learning and Instruction, 4(4)*, 295–312.

Sweller, J., van Merriënboer, J. J. G., & Paas, F. G. W. C. (1998). Cognitive architecture and instructional design. *Educational Psychology Review, 10(3)*, 251–296.

Tabbers, H., Martens, R. & van Merrienboer, J. J. G. (2004). Multimedia instructions and cognitive load theory: Effects of modality and cueing. *British Journal of Educational Psychology, 74(1)*, 71–81.

Thomas, R. & Neilson, I. (1995). Harnessing Simulations in the service of education: the interact simulation environment. *Computers and Education, 25(1-2)*, 21–29.

Tversky, B., Morrison, J. B. & Betrancourt, M. (2002). Animation: Does it facilitate learning? *International Journal of Human-Computer Studies, 57(4)*, 247–262.

Veerman, A.L., Andriessen, J.E.B. & Kanselaar, G. (2000). Learning through synchronous electronic discussion. *Computers & Educations, 34 (3–4)*, 269–290.

von Glasersfeld, E. (1988). The reluctance to change a way of thinking. *The Irish Journal of Psychology, 9(1)*, 83–90.

Wallace, R. H. & Carpenter, S. R. (2005). Using Spreadsheet Software in a "Value-Added" Review of Infrared Spectroscopy in the Organic Chemistry Laboratory Course. *The Chemical Educator, 10(4)*, 308–309.

Wong, S., Popovich, N. D. & Coldiron, S. J. (2001). A Simple Computer-Interfaced Calorimeter: Application to the Determination of the Heat of Formation of Magnesium Oxide. *Journal Chemical Education, 78(6)*, 798–782.

Wouters, P., Paas, F. & van Merriënboer, J. J. G. (2008). How to Optimize Learning From Animated Models: A Review of Guidelines Based on Cognitive Load. *Review of Educational Research, 78(3)*, 645–675.

Wouters, P., Paas, F. & van Merriënboer, J. J. G. (2009). Observational learning from animated models: Effects of modality and reflection on transfer. *Contemporary Educational Psychology, 34(1)*, 1–8.

In: Computer-Assisted Teaching: New Developments
Editors: B. A. Morris, G. M. Ferguson, pp. 37-61
ISBN: 978-1-60876-855-4

Chapter 2

COMPUTER-ASSISTED TEACHING AND CONCEPT LEARNING IN SCIENCE: THE IMPORTANCE OF DESIGNING RESOURCES FROM A PEDAGOGIC MODEL

Keith S. Taber [1*]

[1]Faculty of Education, University of Cambridge, UK, University Senior Lecturer in Science Education, Science Education Centre, University of Cambridge Faculty of Education, 184 Hills Road, Cambridge CB2 8PQ, United Kingdom.

ABSTRACT

Computers and related information and communication technologies (ICT) are increasingly being employed in teaching at school and college level. Like any new educational technology, computers have strengths and limitations; and it is important that computers are used to support existing educational aims when they are appropriate tools, rather than simply being adopted to do what they are good at because ICT is seen as intrinsically 'good'. In science teaching, computers offer a number of useful properties, both related to their use in data collection and analysis, and their ability to offer high quality simulations. These modes of use offer much to support the classroom teacher. However, computers are increasingly being seen as suitable tools for 'delivering learning' in individual study as a supplement to, or even an alternative to, the teacher. Whilst this offers more flexibility - in where and when studying can occur - resources intended to 'teach' learners when the teacher is not present are a different proposition to resources provided as tools for the teacher to employ flexibly to support their own teaching. Conceptual learning in science is well recognised as often being problematic even with motivated students and skilled teachers. This chapter considers the challenge of producing materials to support conceptual learning in science, and the type of pedagogic models that are needed for successful computer-assisted teaching of science. As an example, the chapter discusses the pedagogic model employed in developing learning

* Corresponding author: E-mail: kst24@cam.ac.uk, http://www.educ.cam.ac.uk/staff/taber.html, http://people.pwf.cam.ac. uk/kst24/

material in physics for use in further education colleges in England, and the responses of students to the resources.

Keywords: Resource-design; interactivity; conceptual change; science teaching; scaffolding learning; instructional model; constructivism.

INTRODUCTION

Computers and related information and communication technologies (ICT) are increasingly being employed in teaching at school and college level. Where computers may have been seen as something of a novelty some decades ago, ICT is now seen as a normal part of the resources needed for a properly equipped classroom in much of the 'developed' world (Becta, 2008; Ofsted, 2009). Data projectors are now the norm in many classrooms, as are 'interactive' whiteboards. Computers for student use are common in teaching rooms; and so are suites of machines - for teaching staff to book, or as part of the provision of independent learning facilities.

Like any new educational technology, computers have strengths and limitations, and those that have an interest in the sale and adoption of such equipment clearly wish to focus on the strengths. Without in any way undermining such strengths, it is important to recognise potential issues about the increasing use of ICT in classrooms.

One such issue is that of the nature of what computers are good at, in relation to what classroom teachers need to support them in their work. There is a danger of focusing upon the strengths of computers (e.g. their processing speed; their storage capacity; their networking potential) and developing applications around these strengths, which then need to be 'sold' to schools without consideration of the kinds of tools teachers actually want. This is perhaps understandable and maybe even inevitable, but has potential to distort teaching. If substantial investment is made in new 'kit' and associated applications – designed to fit the strengths of the hardware – there will be pressure on teachers to use the resource: even if it may not do exactly what they need. This is not such an unlikely scenario when the pressure to include and adopt new technology comes from national government-led initiatives rather than teachers themselves (Becta, 2009a, 2009b). An obvious need here is for end-users to be very much involved in the design of educational resources (whether hardware or software). However, even when this happens, there is a likelihood of enthusiasts among teaching staff getting involved, who may not be typical of their colleagues. Enthusiasts will tend to have not only a high level of commitment, but also higher levels of skills and knowledge; and - most importantly - the confidence to be prepared to persevere with an innovation in the face of initial setbacks.

Ultimately, teaching resources should allow teachers to do better the things that they feel meet educational aims and objectives, rather than just offer a good way of doing something simply because it is possible with the technology.

This seems to have been a very real issue in the early mass introduction of computers into schools: with equipment in many classrooms being seldom used because there was little that teachers wanted to do with it (or at least because the things it could do *that teachers wanted to do* needed machines to be employed in modes that were not viable with the ratio of machines to pupils in a classroom). However, it should be recognised that as applications have become available which *do* support what teachers want to do, this has become much less of an issue.

A related problem is the 'cost-benefit' aspect of teachers learning to use new resources. A busy teacher working with challenging classes may well see potential value in doing things differently. But learning new teaching approaches – in terms of learning to use the technology well enough to adopt it whilst multi-tasking in a busy classroom; in terms of teaching the pupils the skills involved; but just as much in terms of adapting pedagogy to best apply the new approach – takes time and effort. Moreover, it is almost inevitable that there will be a learning-and-practice period where things will not be done as well using the unfamiliar approach as they were by more traditional means. Teachers have to not only believe that change is ultimately worthwhile, and should be undertaken 'at some point' (often an easy commitment), but also that they have the capacity to take on the challenge 'now'. In practice, 'now' seems to always be a busy time for teachers, and there is usually a perceived 'easier' time somewhere on the temporal horizon. Professional development support is obviously a key requirement here (Davis & Varma, 2008; Warwick & Kershner, 2008).

Use of ICT in the Classroom

A key use of computers in many classes is to access the Internet. In some ways this is an excellent exemplar of what ICT does well. Traditionally classrooms may have had some reference books - in my own school teaching days I maintained a modest library of science books in my teaching lab - but this is inevitably a limited resource. The school or college library offers more options. However, referring to library books during a lesson poses logistic problems. Time spent moving back and forth between classroom and library is not efficient. In many schools, allowing pupils to freely move around school in this way unsupervised would be frowned upon or even forbidden anyway. Yet even aside from these concerns, the school or college library is likely to have a limited range of up-to-date titles on any particular topic.

The Internet offers an ideal step forward in this regard. Through the Worldwide web (www), any classroom computer can access information on just about any topic; and including the most up-to-date postings. (At least, to the extent that the school computer office does not restrict access to resources – a common complaint of UK teachers is that restrictions meant to prevent connection to sites deemed inappropriate for school children often also act as barriers to ready access of useful resources.) This facility has rightly been adopted in many classrooms, in science as in other curriculum areas. However, as teachers soon realise, whilst computers are very good at accessing information, pupils are often very poor at selecting, critiquing or synthesising sources. The Internet offers a vast – indeed effectively unlimited - resource provided that the discriminating user can apply quality control. Teachers soon find that whilst Internet access is a very good way to keep pupils occupied, and, sometimes, even on-task, it is only an efficient use of student time once some important higher level skills are acquired. This certainly does not suggest the Internet should not be readily available to pupils: many have it at home and are accustomed to being able to search for information at whim (and it is important that resources at school should not be obviously inferior to those at home); and without doubt the critical and effective use of such ICT is a major life skill that all school leavers should have acquired. However, it shows there is a cost to effective use of the resources that must be paid to get the most from the facility.

Another innovation that has been widely adopted is the use of the 'data projector', often in conjunction with an interactive whiteboard (IWB). The data projector allows the teacher to prepare material on the computer and then present it in class. This could be seen as a digital update on the overhead projector (OHP), an earlier innovation in educational communications technology that allowed the teacher to bring to class pre-prepared acetate 'slides' to show during the lesson. As always, the technology may be used in conservative or innovative ways.

A chalkboard allows a teacher to write out their notes for students to copy, rather than having to read them aloud. Yet a teacher who simply came into class and wrote out their notes on the board would hardly be considered to be teaching in the usual understanding of the term – although generations of university students might well recognise this form of 'educational' technique! The good teacher used the chalkboard in a more *interactive* manner: in the right hands 'chalk and talk' can be effective and engaging.

The OHP could also be used just as a 'high-tech' means of notes-transfer: but also had new potential. Having pre-prepared notes on slides makes life in the classroom easier for teachers, but offers little advantage over a chalkboard to a pupil simply asked to copy down the same information as could be written on a board. However, the OHP offers new affordances. A detailed diagram that would take considerable time to draw on the board could be prepared on several layers of acetate sheeting, allowing an explanation to be 'built up' with the displayed figure itself. Overlays can also be used to present incomplete information to test student understanding (cf. the modified 'cloze' procedure commonly used in teacher hand-outs and student textbooks), and material from textual sources that were too small to be seen from all around the class could be photocopied to acetate for projection so that they were visible to all the pupils at once.

The data projector in turn could be seen as little more than a digital version of the OHP, and even in this regard it offers some advantages – neater text; ability to include diagrams and photographs sourced from the internet etc.; ability to reveal information point by point so as not to overload pupils. (Of course seasoned OHP users achieved much the same effect with a paper sheet being moved down the acetate as it was revealed – but the data projector does not suffer in the same way from the potential effects of draughts.) Yet the real potential of new technology, though, must be in *enhancing pedagogy*. Where IWB are simply used as display screens for data projectors, they could as well be effectively and cheaply replaced by the application of some matt white paint. Where the real interactivity of the IWB is used, teachers are able to explore the potential for enhanced pedagogy that new technology offers. New technology actually supports innovation *not* by being novel kit, but by enabling novel application. In the case of IWB the potential would seem to lie in developing the dialogic aspect of teaching and learning (Hennessy, Deaney, Ruthven, & Winterbottom, 2007; Warwick & Kershner, 2008), something that has been recognised as key to student learning in all areas, including science education (Mercer, 2004).

Using Technology to Teach Science

It has been claimed that "despite widespread use of technology by scientists across many disciplines, computers and network technologies are often underutilized and poorly integrated into core science education activities" (Songer, 2007). This is a great shame, as in the science classroom new technology has offered a number of especially appropriate resources for teaching. One of these areas is data-logging: using computers to collect and store experimental data.

There is immediately an issue that should be addressed here, of possibly excluding pupils from essential aspects of the scientific process. Making observations and recording measurements is a part of much empirical work in science. In many areas of science these processes are often now automated: computers collect the data, store ('log') it, and then offer various ways of displaying it. Indeed, often the raw data is not actually examined by a human being at any stage; as the data is 'treated', or even fully analysed, before being presented to the scientists themselves. This creates a question about whether we want science education to be authentic – like working in real modern laboratories – or to be about understanding principles and processes, in which case it may be a mistake to have machines undertake key stages in data collection and analysis. Of course, this is not *really* an 'either/or' question: we want students to both understand the processes and experience some degree of authenticity. The question then is *when* it makes good educational sense to use technology to automate school science practicals.

An immediate answer is: not until pupils have had some experience of simple practicals where they can be involved in all stages of the data collection and analysis processes. It may make sense for pupils to undertake a relatively straightforward practical 'by hand', before repeating effectively the 'same' practical using a data-logger (or data-logging computer). This would allow the pupils to best appreciate precisely what the data-logger is doing, and also ensure enough familiarity with the specific science and practical to allow them to focus primarily on the data-logger as a tool they can use. Some of the issues here reflect the debate in mathematics education around if and when pupils should use electronic calculators rather than calculate in their heads (and no doubt much the same debate was previously held in the era of slide rules and log tables).

Once students are used to the data-logger, it can then be used in subsequent lessons if and when it offers a specific advantage. Examples would be when observations need to be made over extended periods (e.g. set-up in a lesson, and left overnight), or with higher frequency than pupils could achieve (for example with fast changes), or when several probes are used to take simultaneous readings etc. The possibility of watching graphs plotted in real time, *and* being able to adjust axes as becomes necessary, also offers an obvious advantage. Having to discard a graph drawn by hand part way through because data moves outside the expected range can be an interesting learning experience, but may also be quite demoralising to some pupils.

Once students are familiar with data loggers they can be incorporated into genuine student enquiry tasks to offer learners an authentic experience of what it is like to do science (Davies, Sprague, & New, 2008). Data loggers can also be used in teacher demonstrations with great effect: such as using sensors to convert a pupil's movements in the classroom into distance-time or speed-time graphs. Other related applications include the use of digital cameras and microscopes that allow the teacher to share images with the whole class in new ways.

THE COMPUTER AS A TEACHING MACHINE

The applications discussed so far relate to the teacher using the computer as an instrument or tool to support their teaching, in similar ways to other classroom equipment such as the OHP, or traditional laboratory apparatus such as Bunsen burners and microscopes. These applications make use of the speed and capacity of modern machines and their ability to interface with information sources such as the www and other devices such as temperature probes and cameras,

However, a more interesting aspect of ICT development is the possibility of making more use of the interactivity that can be built into computer software: that is in the area of teaching machines. The notion of a teaching machine is not new, and indeed programmed learning is an approach that need not rely on electronic computers (Bugelski, 1971).

It is possible to design programmed learning simply as a set of numbered cards. Each card can present some information, and a question with several possible answers. The student selects an answer. Depending upon which answer they chose, they are directed to a different card. (The same principle is used in some surveys: e.g. if you answered yes to question 7, please ignore questions 8-11 and move on to question 12.) A correct answer is taken to show understanding of one key point and the student is moved on to the next teaching point. If distractors (incorrect response options offered) are well designed they link to specific likely misunderstandings, and so the student making incorrect responses is directed to cards which respond to those specific points, before returning the student to the original card (or a parallel one covering the same core information).

The logic behind such materials is quite well motivated. Most school age pupils are not very skilled in learning from texts, so just reading a chapter in a book is not an effective way to learn. When teachers teach, they intersperse their informative statements with interrogative ones. They ask questions that test student understanding, to inform decisions about what to say next: move on, repeat, go back over prerequisite material etc. Programmed learning materials are an attempt to build into a text the kind of interactivity that comes from working with a teacher or tutor. One set of well-designed programmed learning materials can potentially offer effective individualised routes through a topic for different students. The gifted and knowledgable move through quickly without inefficient and de-motivating wasting of time; the lower attaining student gets plenty of support and feedback, and several exposures to the key ideas on their path through the materials. When expressed like this, it might seem that such materials offer an advantage in a classroom situation: no matter how great the skills of a teacher, class teaching is always something of a compromise when trying to effectively and efficiently simultaneously teach the inevitably heterogeneous learners in any class. No matter what policies and procedures may be used to band or set students in a group, in practice every class is to some extent a 'mixed-ability' class: with each learner having their own strengths and areas of prior knowledge.

However, it is also clear that this advantage of programmed learning materials can only be realised if the materials themselves are well designed. Indeed, in any substantive area of knowledge, programmed learning materials to meet the needs of a range of students would need extensive research and design, and are likely to become complicated and extensive. Certainly this would be the case if the materials are to come close to offering the level of interactivity available from a human teacher. The effective teacher calls upon three major knowledge domains: of the subject being taught, of the relevant subject-specific pedagogy, and of the pupils in a particular class. Computers, however, only 'know' what is programmed into them, and in most science teaching contexts it is currently unrealistic to think that it is possible to programme them to be the experts we expect teachers to be.

An important exception is in using computers to simulate virtual environments. Computers can be programmed with rules representing physical laws, and can quickly calculate – for example – how a projectile would move from certain initial conditions given particular physical laws and constant values. This is an example of a type of application where the processing power of the tool offers an ideal resource that can (within its own range of application) be highly interactive as students experiment with changing initial conditions, or coefficients. Indeed working with such virtual micro worlds offers a real kind of experiential learning (Heuer & Blaschke, 2001; White, 1993).

In other science study contexts - learning about say the digestive system, the reactions of acids, or electromagnetic radiation - computer-based learning is far from offering the interactivity available from a human teacher. However, despite the challenge of producing effective programmed learning materials, what is clear is that electronic computers – the epitome of the programmable machine – offer the ideal hardware on which to run such materials. Whilst teaching machines that can replace classroom teachers may well still be some way off – if indeed such a notion was considered desirable – there is much potential for at least incorporating some of the basic features of programmed learning into software designed for educational use.

CHALLENGES OF CONCEPTUAL LEARNING IN SCIENCE

However, science teaching offers many challenges to the human teacher, and these need to be considered by those designing educational software that is intended for any degree of independent use (i.e. other than under direct supervision of the teacher). Indeed, the principles here are probably common to most 'academic' learning, but have been widely explored in the context of science learning, where it is known that students commonly come to class with their own 'versions' of science concepts at odds with the authorised versions presented in the curriculum (Duit, 2007). As children construct new knowledge on the basis of what they already understand and (think they) know (Ausubel, 2000; Taber, 2009a), these ideas are very significant for learning.

Indeed there has been a vast research effort that has explored the ideas about science topics that students bring to class; the nature of those ideas; and how best to shift student thinking towards the target knowledge set out in official curriculum documents (Taber, 2009a). This has shown that in virtually any science class – at any level, working on any topic – students are likely to hold 'alternative conceptions'. It is well recognised among science educators that many such conceptions have consequences for the intended learning (Driver & Erickson, 1983; Gilbert & Watts, 1983).

Studies suggest that the alternative ideas that students bring to class vary considerably along a range of dimensions: but that in some cases they may be strongly committed, tenacious, well-integrated into the student's frameworks of knowledge, and applied widely across contexts (Taber, 2009a). In some case such ideas have been found to be highly resistant to 'correction' during to teaching, and likely to reappear in student thinking in the medium term even when end-of-topic testing suggests short-term shifts towards the scientific models presented in the curriculum. A number of theoretical perspectives seem to usefully inform our understanding of the nature and development of at least some of these alternative ideas (Chi, 1992; diSessa, 1993; García Franco & Taber, 2006; Karmiloff-Smith, 1996), and research has suggested various strategies for responding to different classes of alternative conception (Bryce & MacMillan, 2005; Driver & Oldham, 1986; Treagust & Duit, 2008).

However, the literature is substantial, and even after some decades of research, it is still not possible to offer teachers clear and effective generalised guidelines for responding to students' alternative conceptions in science: even if it is now clearer how the field needs to move forward (Taber, 2009a). What seems clear is that whilst there are certainly some very common alternative ideas that will be found amongst learners in most classes (Taber, 1998; Watts & Zylbersztajn, 1981), many ideas students have are idiosyncratic (Taber, 1995) and need to be identified on an individual basis. So despite the undoubted value of research in informing teachers about the scope and nature of learners' idea in different topics, and different teaching approaches that can be employed, successful teachers need to coordinate this knowledge with their own specific understanding of particular classes and individual pupils. Teachers need to act as 'learning doctors' who diagnose student thinking at the individual level, and respond to them accordingly (Taber, 2001, 2002).

Developing Computer-Aided Learning from a Pedagogical Model

This need to consider the individual ideas of particular students, and how these existing ideas will shape new learning, makes effective science teaching a highly interactive process. (And to the extent that children bring alternative conceptions to class in other curriculum subjects, much the same will apply to these subjects.)

This consideration adds complexity to any attempt to develop effective learning resources, whether they are textbooks, laboratory apparatus, or software. However, most learning resources bought by schools are intended for use under close teacher supervision. Yet interactive software offers the promise of being something that can support learning with minimal teacher involvement: something learners can use in the library, or at home, freeing up the teacher to focus on other students, or allowing the teacher more time to spend on other topics or activities.

Such software needs to be a lot more than a textbook transferred to a computer, but ideally will provide at least the following features:

- providing the learner with an overview of the area to be covered near the start;
- providing navigation options to allow the learner to move about the material according to their own needs;
- offering questions with feedback for the learner to check their understanding;
- 'scaffolding' to allow learners to build up their understanding at their own pace;

Software which looks to respond to such requirements will be considered here as encompassing an instructional model.

Conceptual Learning and Scaffolding

As suggested earlier, a key feature of conceptual learning is that what is learnt depends upon what is already known: existing knowledge provides the conceptual frameworks through which new information is interpreted and understood. This is why alternative conceptions can be so significant, as they distort the teacher's meaning so that the learner unknowingly understands material differently from intended. Learning is often seen as a process of knowledge 'construction' with existing understanding providing the foundations for new learning (Taber, 2009a).

Research in the cognitive sciences further shows that this 'construction' process is constrained by aspects of the available cognitive apparatus, as well as the conceptual resources available (Miller, 1968). Students learn science, or indeed other subjects, incrementally in terms of limited 'learning quanta' (Taber, 2005), even if there may be critical points where the accumulated material allows apparently sudden shifts in understanding (Gilbert & Watts, 1983). Designing instruction needs to consider both the 'building materials' students already have available, and the construction 'tools' they can use (Taber, 2008).

A key concept here is that of 'scaffolding', a notion developed out of the work of the psychologist Vygotsky by Jerome Bruner and his colleagues (Wood, 1988). Scaffolding is commonly described in terms of how teachers provide vicarious support for a student building up a new skill or conceptual framework whilst they are mastering new learning, and then gradually 'fade' that support as the student is able to take over. In terms of designing learning resources that can achieve this, the design has to provide 'scaffolding POLeS' (Provided Outlines Lending Support). However, often before students can take advantage of such devices, they need to first bring to mind and organise their existing relevant knowledge. These earlier types of supports – 'scaffolding PlaNKs', Platforms for New Knowledge – are also important in facilitating effective learning (Taber, 2002) – reflecting an important feature of so-called 'advance organisers' which can support meaningful learning (Kember, 1991).

In terms of educational software, the instructional design has to help the students recognise which prerequisite knowledge is relevant, and show them how it fits with new knowledge – as well as offer a way of structuring prior learning with new information to build up new understandings. To do this well is very demanding – after all when teachers scaffold learning in this way they are constantly interacting with learners and acquiring feedback that allows them to adapt their input.

To explore something of the demands of adopting these features into ICT based resources, one example of government commissioned educational software will be discussed in some detail.

The Epic Design for Computer-Based Learning Resources

The example discussed here derives from some independent learning materials produced in the UK for the further education sector. The school leaving age in the UK is 16, although all school leavers are entitled to two further years of free education or training. Colleges that offer courses for those who have completed school, and which are (mostly) sub-university level, are considered to be 'further' education (FE) colleges. University level education, at Honours degree level or above, is labelled as 'higher' education (HE). FE colleges may offer a very eclectic range of courses to meet the needs of a wide variety of full-time, part-time and evening class students: those studying for entry to higher education, as well as basic skills (e.g. literacy and numeracy for adults), trade skills (motor vehicle, building trade etc), and professional qualifications (such as the professional accountancy and insurance examinations). Some FE colleges also teach parts of HE courses under the guidance of Universities, and they may also teach some students who are in the last few years of compulsory education (usually those who have been excluded from, or refused to attend, schools, and where although further school attendance is considered non-viable, the students are judged likely to respond to the more adult environment of a college). The FE sector also includes the so-called 'sixth form' colleges who specialise in teaching just 16-19 year olds, and often focus on courses leading to university entrance examinations.

The example of a computer-based learning resource considered here derives from teaching materials prepared to support students working at this level, i.e. working towards the 'A level' qualification. Students commonly study four A level subjects, and places for university courses are commonly awarded conditional upon certain minimal grades (with specific subjects required for many courses).

As part of a government funded initiative, a public sector organisation called the National Learning Network (NLN) commissioned the production of independent learning materials to support students working towards A level examinations in colleges in the FE sector. The contract for developing materials for physics (and several other subjects) was awarded to a company called Epic (Epic Group plc). Epic designed eight units of learning materials across a range of A level physics topics (Epic, 2004):

- rectilinear motion
- radioactivity
- electricity
- resistor
- circular motion and oscillations
- waves
- quantum phenomena
- fields and forces (two modules)

Within each unit there were several (normally two) different 'learning objects' (LO), i.e. specific foci for student learning (see Table 1).

The modules were drafted by 'subject matter experts' (experienced teachers of Physics at Advanced level), according to an outline 'template' designed by Epic and informed by their instructional model (detailed below). [1] Materials were trialed with focus groups and in

[1] The present author was involved in the development of the NLN Physics materials on a consultancy basis to Epic, as one of the subject matter experts.

authentic teaching contexts before the materials were delivered to the NLN to be made available to colleges.

Table 1. Topics included in (UK) National Learning Network Level 3 Physics blended learning materials.

Unit	Learning object (LO)
Rectilinear motion	Conservation of linear momentum Measuring the acceleration of free fall
Radioactivity	Properties of alpha, beta and gamma radiation Experimental determination of half-life
Electricity	Conductivity and resistivity Resistance change of a thermistor and a light-dependent resistor
Circular motion and oscillations	The motion of a simple pendulum Demonstration of forced, free and damped vibrations
Waves	Polarisation of transverse waves Stationary waves Diffraction of water waves and light waves
Quantum phenomena	Demonstration of the photoelectric effect Demonstration of the electron diffraction phenomenon
Fields and forces (1)	Newton's universal law of gravitational attraction The gravitational field strength at different distances from the Earth's surface Verification of Coulomb's law Electrical field lines in uniform fields
Fields and forces (2)	Principles of a cyclotron Principles of different radiation detectors

The materials were expected to be suitable for use in a range of modes. Of particular relevance here is their use as independent learning materials that could be accessed, in a library or resource area, without supervision by subject specialist staff. However, the materials were not intended to replace normal classroom based learning, and also had to be suitable for other modes, such as acting as taster or remedial materials, or being used with a class led by a subject specialist teacher.

The Instructional Model Used in the Epic Materials

The instructional model adopted for the NLN Physics materials involved a sequence of stages, each realised as one or more screens that the student would met in a sequence if they worked through the materials in the recommended order.

Each unit began with an introductory screen that was designed to engage the learner's interest and introduce the topic covered – i.e. an orientation phase. Next comes a screen concerning the aims of the unit, setting out what the student should expect to learn when completing the unit.

This is followed by a screen eliciting 'first thoughts' - where a student is asked a question to get them to think about their existing understanding of the topic area. (This screen could be considered to act as a scaffolding 'PlaNK' in the terminology introduced above.)

The next screen or two make a 'presentation' of the main ideas being explored in the unit. This usually involves some form of interactive animation, and was intended to be suitable for being used in 'demonstration' mode by a class teacher as well as being part of the sequence met during independent learning. For example, the student interacts with an animation to make changes in a circuit and observes the change in the reading on an ammeter; or the student moves different materials between a radioactive source and a detector, and observes the change in the rate at which decay events are detected. There then follow a number (c.5) of 'investigation' screens, which allow the learner to explore the key idea in more detail (and act as scaffolding PoLES in the terminology used above).

After working through the 'presentation' and 'investigation' sections offering the core knowledge, the learner is invited to apply their knowledge by answering an open question that tests understanding of the learning objectives built into the unit. A model answer is provided that the student can compare against his or her own response. This is followed by a 'summary' which highlights the key learning points, and then a 'check your understanding' section of three screens offering objective questions (where the student responses can be logged for a tutor).

The instructional model is illustrated in Table 2. The learning materials were intended for use in a virtual learning environment (VLE), and included features such as colour screens with illustrations, links to glossary entries, and audio versions of text (which were tested at the UK Royal National College for the Blind).

A Case Study of the NLN Physics Materials Being Used in Independent Study

Once the EPIC materials had been completed and prepared for trials, I asked several students studying physics at A level if they would work through a topic for me, so that I could get a feel for how well the instructional model supported independent student study.

Given the limited space here, I will briefly discuss how one student, who I will refer to by the assumed name of Adrian, got on with some of the materials. Adrian worked through two of the units, during separate sessions. On each occasion Adrian worked alone at a computer on trial versions of the software, observed by the present author.

Adrian was instructed to work in a 'think aloud' mode. 'Think aloud' or 'talk aloud' is a technique used by researchers investigating how people think when carrying out a task (Phang, 2009). The research participant is asked to vocalise their thoughts so that the investigator is able to follow the participant's reasoning.

As with all research techniques, there are disadvantages (Branch, 2000). Firstly, the technique is clearly limited to the participant's conscious thinking: even an honest and full report does not give access to implicit thinking. This is not a trivial point, as we are not always aware of all the stages in our thinking, and sometimes the reasons we give for certain decisions may be (without our realising) after-the-event rationalisations of our sub-conscious decision making (Taber, 2008). Secondly, conscious thinking is not all verbal in nature. So for example, Einstein reported that much of his thinking about physics was in images (Miller, 1986): which would have to be re-represented by being described in words to be reported. The very act of 'translating' thinking to verbal form, and so unavoidably and inadvertently modifying it, is likely to change the course of that thinking compared with how the individual's thinking would have proceeded without this step. Thirdly, there is clearly a cognitive load on asking someone to report their thinking out loud: such an additional demand surely slows, even if it does not impede, the normal thinking processes. These demands are such that researchers who use talk-aloud regularly report that a small-but-significant minority of individuals prove unsuitable as participants in such research – being unable to both proceed with the task *and* provide a full running commentary at the same time.

Adrian was a sixth-former (i.e. in post-compulsory education), who was studying A level sciences including physics. He was in the second year of his two-year course. Adrian's sixth-form centre was part of a secondary school, but he was in the same age range and undertaking the same course as students studying in FE colleges for whom the materials were designed. (Diversity is a characteristic of the English education system, so 16-19 year olds studying 'A levels' may be in schools, specialised sixth form colleges, or general FE colleges.) Adrian was one of a number of students at his school who had volunteered to participate in a project to explore aspects of student thinking and learning about school science. The *Understanding Science Project* (Taber, 2009b) was based on sequences of interviews exploring student thinking across different science topics they met in schools.

Table 2. The instructional model used by Epic for the NLN Physics materials.

Section	Notes – how the screen(s) in the section fit into the overall design
Introduction	Designed to engage student. Links concepts of topic with some real-life application judged likely to be relevant to students (mostly 16-19 year olds).
Aims	Allows the student to see the scope of the work they will undertake, and what they are expected to achieve – helps act as an advance organiser which focuses the student on key issues/concepts
First thoughts	A question to elicit the student's existing thinking about the topic.
Presentation	Interactive demonstration of the key principle or idea. Offers a simulation of a hands-on investigation of the physics, and gives a visual impression of the key phenomenon.
Investigation	This section fills out the details around the key idea from the previous section.
Application	A question for the student to apply the ideas they have met in the unit, and a model answer to help them judge if they need to review any of the previous screens.
Summary	The ideas met in the unit are set out in a single summary screen to review learning.
Test items	A series of objective questions testing understanding of the ideas met in the unit.

Usually students are only allowed to enroll onto A level courses if they have performed well on the school leaving examinations (the General Certificate of Secondary Education). However, many students find Physics a difficult subject at this level, even when they previously coped well with school science. From my research sessions with him I considered Adrian to be capable and motivated student: genuinely interested in physics, but without demonstrating exceptional ability in the subject. Adrian was invited to try out some of the NLN materials, and he agreed, recognising that this could be a useful additional learning opportunity.

Adrian first worked through the learning object on *the motion of a simple pendulum*, and then a few weeks later the learning object on the *demonstration of the photoelectric effect*. On both occasions he worked in a quiet room at the University with only the author present. Adrian's school was only about a kilometre or so from the University's Faculty of Education, and as he lived very close to the Faculty it was decided to undertake the sessions there, away from potential distractions at school. Both sessions took place in rooms in the Science Education Centre in the Faculty: one in a teaching room, the other in the author's office. The sessions were held in Adrian's free time and did not require any withdrawal from his usual classes.

Here I will briefly discuss the LO on *the motion of a simple pendulum*, and consider how Adrian fared in interacting with the materials. To give a flavour of the procedure, I'll quote from the start of Adrian's think-aloud protocol (as recorded and transcribed by the author) as he start looking at the LO on *the motion of a simple pendulum*. Inverted commas within the extract indicate material that Adrian read aloud from the text on the screen.

> "So it starts with erm, talking to you about history, erm it's got 'the pendulum has played an important role throughout history'. It's then got four pictures, one of an old guy, an old painting, which when I put my erm cursor on it, it comes up with the history of pendulums, the next one, number 2, is erm, just waiting for it to come up, it's not going to let me, oh it's also the history of pendulums, so I'm guessing it works through the erm, like a chronological order. Start with the first one, number 1, 'the pendulum gave us accurate time keeping from 1657 when Christian Huygens invented the pendulum clock'. 2, this is just giving me more history, different names of people, talking about seismographs, and then it goes into simple harmonic motion at the end. Erm. [At this point the author offered clarification on navigation through the screens] So - [pause, c.4s] it's putting the pendulum into the modern day, erm, environment of a TV programme. [Pause, c.7s] And it's saying what the aims of this section of the computer programme will tell us. [Pause 6s] It's good the way which it's got links like [i.e. to the Glossary], so you can link to different words which you might not know the meaning of. ..."

The Introduction to the LO gave, as Adrian readily recognised, some historical background to the science behind the pendulum. As well as Huygens (the 'old guy'), it also presented information about Foucault's pendulum and the use of pendulums in seismographs, before explaining that "the pendulum has simple harmonic motion..." which was "a characteristic of many other oscillating systems and its theory can be applied to electronics and waves". As the quoted protocol above suggests, Adrian passed over this material quickly, apparently considering it as little more than a preamble.

The materials then offered a context for thinking about the use of pendulums: a television game show where "the set is designed with a large pendulum which swings across the studio. The contestants will have to answer the questions correctly while the pendulum is swinging over their half of the studio". The student using the software is told that "you will advise the producer of the show on how he can use the properties of the pendulum to improve the show's format". The use of such contexts is intended to engage students by showing how the science is relevant to contexts likely to be of interest, and Adrian seemed to recognise this common ploy ("it's putting the pendulum into the modern day, erm, environment").

The 'first thoughts' segment of the LO is intended to elicit the student's prior learning about a topic, and to focus their thinking upon important features of the phenomenon. For this LO this comprised of a passage where the student had to respond by selecting the right response at a number of points where several words (*italicized* in the extract below) were offered. Adrian was quite familiar with this kind of task and proceeded:

> " 'When a pendulum swings, it travels fastest as it goes past a central point, which means it's kinetic energy is' and it's giving you a list I think, oh yeah, *constant, maximum, minimum, zero,* so I think it's just trying to find out what sort of things you know. Erm – I'm trying to remember what the answer to the question is, er, as it, when it's at its further point away, it hasn't got any potential energy, maximum potential energy, when it's let go, kinetic energy would be at a *maximum* as it goes through. 'As the pendulum swings away from the central position the magnitude of its displacement' so I've just thought what displacement means, it means distance, erm, as it 'swings away from the central position the displacement' *increases.* Erm, displacement's highlighted, I don't know what that means, I'm sure I could find out. I'll try that. [Adrian refers to glossary] ... Erm, question continues, 'kinetic energy is' – *decreasing.* 'Potential energy is' – *increasing.* 'The something of a pendulum, is the time taken', the *period* of a pendulum. 'The maximum displacement of the pendulum from the central position is the' - *amplitude.* Oh, something happened there. I pressed back by accident. Oh."

Adrian managed to work through the items, and to refer to the glossary to check the meaning of a term, only hitting difficulty when he then clicked in the wrong place after selecting his answers. He then corrected his mistake and proceeded to the next screen, which offered the model answer.

Ideally in a think-aloud study, the researcher sets up the participant with the task, and then retires to observe (perhaps often being out of sight of the participant, monitoring remotely by closed-circuit television). However, I soon became aware that Adrian was tending to pass over or miss aspects of the learning materials, and that some interjections and prompts seemed appropriate.

The next part of the LO concerned the 'Presentation' where an animation allowed the student to simulate changing physical characteristics of a pendulum and find the effect. So, for example, changing the length of a pendulum alters its period; but changing the mass of its bob does not (although it does influence how quickly the oscillations are dampened and the swinging stops); and – to a reasonable approximation – nor does the initial displacement of the bob. This is why pendulums can be used in clocks – as highlighted in the introductory screen – as the period remains unchanged as the pendulum dissipates energy and the amplitude of the swing decreases. In the context used for the LO the student is told that "in order to advise the producer of the TV show, you will need to find out more about the format of the show and the set. This animation shows how the show is affected by using three different lengths of pendulum." The instructions read "Press start to see how the pendulum's length affects the time the contestants have to answer the question".

Adrian: [Truncated reading] Press start to see how the pendulum effects, length effects time.

Keith: Did you get any feedback on what you just did?

Adrian: No, I don't think so. Oh. Okay so you've gotta - It's presentation now. Small pendulum. People are talking. I'm not entirely sure what it's showing me. Oh the effect that length of pendulum has. Okay so it's flashing up equations at me. However, oh, - I'm not sure, they're not linking the equations to – anything.

Keith: So what do you think that's meant to demonstrate then, that bit?

Adrian: The equation's T, square root of one over g, I think, oh, '2π square root 1/g', so it's talking about the time period, but it doesn't – effects the time, the contestants have to answer the - . It's showing that, erm, the longer the pendulum, the longer the time period, I think. However, that's not very clear.

Keith: Right, so is that what the animation does? What does the animation do?

Adrian: Yeah. It says 'Press start to see how the pendulum's length effects the time the contestants have to answer the question below', there is no question below, but, - so a medium pendulum, she gets it just as it goes past her, long pendulum, - she gets it well before it gets to her, so it's showing that it takes longer for a longer pendulum to swing, than a short pendulum.

Space here does not allow a detailed account of Adrian's progress through the LO, but these extracts from the protocol offer a fair impression of his engagement with the materials. Adrian was interested in the materials and seemed keen to work his way through: but he did not always seem to appreciate the intended interactivity of the materials without prompting by the 'observer'.

So Adrian was next asked to undertake a calculation:

"I think there was a question on the last bit; I must have just missed that. [Going back to check:] Okay what energy's involved? 'The pendulum will have blob [sic, bob] with a mass of 5kg. Just before the show starts the pendulum will be pulled to one side of the studio, lifting the blob a vertical distance of 1.2 m. The pendulum's gravitational potential energy will be converted energy as it swings. Click on each item to find out more. [1] How much gravitational [potential] energy does the pendulum have as it starts to swing?' [2] So as I was reading through that I was thinking sort of what formulas do I know, erm, relating mass to erm energy, and distance to energy. [3] 'How much gravitational potential energy does the pendulum have as it starts to swing?' [4] 'Ignoring the effects of friction and air resistance the gain in gravitational potential energy's mgΔh so in this case it is 5kg times 9.8' which is the gravitational field strength constant, or, yeah, 'times by 1.2m' roughly 60. [5] The next question is 'what is the maximum kinetic energy of a [sic, the] pendulum?' "

In this extract Adrian [1] reads out a question and [2] then correctly realises that he needs to identify the appropriate formula needed to undertake the calculation. However, instead of doing this he [3] re-reads the questions, then [4] reveals and reads out the model answer, before then, without any further reflection, [5] moving on to the next question.

Clearly in designing material of this kind, the intention is to get the students thinking, so that they are actively processing the ideas, and *then* offer them feedback on their answers. Adrian however showed no compunction is completely short-circuiting that process. At this point, I intervened to explore why Adrian had not made any attempt to work out the answer before looking at the answer:

Keith: What if you think about [the questions] before you press the buttons, is that possible?

Adrian: Yeah, that's what it's asking.

Keith: Or is that something you would tend not to do, because – you know – there's a button that gives you the answer?

Adrian: [Laughing] I guess you might think about it, but if you're using this as an initial learning aid, you wouldn't have a clue where to start, er, erm. It doesn't give you any hint where to start. Like I could understand if it gave you the formula and then said work it out, or erm, these things, these specific things erm will give you this or something if you use them in a certain way, but it's just asking you a question which, unless you've done work on, erm, on that topic before, you probably wouldn't be able to answer.

Keith: I see, what was the first one, what did the first one say?

Adrian: The first question was 'how much gravitational potential energy does the pendulum have, at the start of the swing?'

Keith: So what would you need to know to answer that?

Adrian: You'd need to know the relationship between kinetic energy and potential energy, how if you've got maximum potential energy in a pendulum there's no kinetic energy because it's at its highest point. You'd need to know an energy formula to work our gravitational potential energy. Erm.

Adrian suggests two components of prerequisite knowledge here, but actually only the latter is needed. (Even if it was not clear to a student that the pendulum had no kinetic energy at the highest point, this information was not needed to calculate the potential energy.) The formula for gravitational potential energy was something Adrian should have known, and would likely have used at various points on his course.

Keith: Would you know that already?

Adrian: Not if you were using this for the first time, or if you'd had no other contact with, erm, if I'd just walked in first physics lesson on pendulums and, erm, circular motion and simple harmonic motion and everything, erm, then you might not know it.

Keith: So if the teacher said we're doing a new topic and this is about pendulums, go away and look at this software, [yeah] then you wouldn't think 'I know how to answer that question'?

Adrian: No, if you spent a lesson going through pendulums, and the effect pulling further away has on the energies, which would get you thinking about, well, do I know the formula for potential energy at a certain height or something?

Keith: Do you know that?

Adrian: I probably would have known it from like GCSE [secondary school work].

Keith: Not from pendulum work necessarily?

Adrian: No, however you've got to be able to like relate it, and that's, although you may know the formulas [sic] it's being able to think 'right, that's a formula I know', cause like in circles, there's different formulas for acceleration, acceleration's v-r-squared over something, or something like that. … whereas in an, just in a normal erm, like pulling something along a surface, f = m a, acceleration equals force over mass, and so you've got to be able to sort of see the problem, almost…

The calculation that Adrian was being asked to undertake required the application of a relationship that was considered by the software designers to have been prerequisite knowledge for the topic. It was expected that most students working on these materials would be able to access and apply this knowledge in the new context. At the end of the exchange above, Adrian acknowledges that he should have been able to tackle the question - as he had previously met the formula, and identifying which formulae apply to different contexts was a skill expected in physics.

Presumably if a physics teacher introducing the topic of pendulums had asked the same question, Adrian would have assumed that the teacher had good reason to expect him to be able to respond to the question. Yet in the context of working with the software, Adrian simply assumed he would not know the answer without completing the work – even though it was presented early in the teaching sequence. Adrian did not seem to consider that the materials would have been carefully designed following an instructional model.

The session continued in this fashion, with Adrian working through the rest of the LO, but with regular prompts and questions from the author. The process took just over an hour.

When Adrian had completed the LO, I asked him about his overall impression of working through the software. At this point, having completed the full sequence, Adrian now seemed to recognise how the LO had been designed to follow an instructional model:

> "It works through a process that you will have worked through in class. Like that's the same type of process I worked through when I did pendulums. ... It's er, it's similar in that you get given an idea of what it's going to be about in the beginning in the aims, and so it's sort of erm gets you thinking about possible things that can be linked, erm and then it starts off with things which you will probably, well should already know, like the initial energy equations, erm and it sort of tells you that they are related to this topic, it then gets more involved with a new equation that you probably haven't seen before, time period, and the effect the time period has on, or how you work out time period from graphs and stuff. Erm, what does it do then? Oh yeah, and then it uses that equation, so it makes sure you can use it, erm it didn't just throw you straight in with a four-part question, starting with energy, talking about time periods, simple harmonic motion getting completely confused, it gave you a simple question. Erm, It then, what did it do then? Oh yeah, it then asked you the three questions, the first one was using the equation you had just learnt, again slightly more complex, because it was a different idea. The next one was erm using almost everything you had done in this little section, it's talking about energy and the equation. And the last one was general ideas that you need to be able to remember when you are answering questions, so the true and false. I think the way in which it worked through was sort of building everything up, so that when you go into an exam, you think about the right things."

Adrian had found the experience useful in reviewing the topic area, although he claimed not to find the use of a familiar context (the television game show) especially engaging,

> "I don't think it would have bothered me if they had [just] given you a diagram of a pendulum swinging. I don't think that would have made any difference to me, the game show. I guess they're trying to make it more related to things, but I mean I don't think it makes a huge difference, once you get to like upper sixth physics, you know what I mean? ... I mean maybe like the year 7, 8, 9s [i.e. 11-14 year olds], then yeah, they probably don't want to be doing physics half of them, and so like linking it to something else, taking their mind off the actual physics is a good idea"

It is important not to read too much significance into a single case of a student using learning resources, especially in a format that was unfamiliar. However, although Adrian found the materials useful, it seemed he would readily have ignored much of the potential of the interactivity of the resource, and simply skipped through to where he was being given information, had be not been guided by the author.

A few weeks later Adrian undertook a second LO (*demonstration of the photoelectric effect*), and at the end of that session he was very positive about the resource: "I thought that was really good actually...they sort of talk you through, building it up, ... and I felt I understood it because of they way they built it up, and they told you specifically there were three things that, three factors, things like that". When asked if there was anything about this LO that he did not find helpful, Adrian suggested "no, I think it was all necessary, like, to make sure you understood it. It was all necessary stuff". Presumably his earlier experience with the first LO had helped prepare him to better use the interactivity of the resource.

The overall message from this particular case is that the NLN Physics materials were based around an instruction design which had strong potential to engage students in the learning process; but that students may be tempted to bypass the built-in interactivity – especially if they did not expect (perhaps based on their experience of using other educational software) that the LO would be carefully designed as a learning resource.

An Example of the EPIC NLN Physics Materials Being Used in Supervised Study

The second use of the NLN materials discussed here was with a group that would not be considered part of the intended target audience for the materials. These were secondary level students, 14-15 year old pupils, attending an enrichment programme called ASCEND. ASCEND, *Able Scientists Collectively Exploring New Demands*, was an after-school programme designed to meet the needs of the most able in secondary science classes (Taber & Riga, 2006). ASCEND was supported by funding from an Educational Charity, the Gatsby Science Enhancement Programme, and was based at the Faculty of Education in Cambridge, working with Comprehensive Schools in the City of Cambridge (Taber, 2007a). The four partner schools identified students in Y10 (14-15 year olds) who they felt would benefit from being challenged in science, and were interested enough in the subject to want to attend after school sessions at the University. This was seen as one way to support schools in meeting their expectations under government 'Gifted and Talented' policies (Taber & Riga, 2007). However, as the notion of giftedness in science is a disputed concept, and the approaches adopted by schools to identifying students for schools 'gifted' registers can seldom be considered rigorous or sophisticated (Taber, 2007b), there was no strict requirement that all delegates (attending pupils) should be formally included on the schools' list of students gifted in science. Nonetheless, the 30 or so delegates who attended sessions were certainly among the more able in their age group, and some would certainly be considered gifted by most criteria.

ASCEND sessions were each organised around different themes, most of which were linked to the nature of science (demarcation, explanations in science, philosophies and scientific method, scientific laws, models). However for one session students were given the option of working with the NLN physics materials, and most of the delegates chose to take up this option.

The NLN materials were designed for older students who had successfully completed their school level science, so it was clearly a risk to invite younger students likely to be lacking some of the assumed prerequisite knowledge to use the materials to try out the programmes. Five of the LO which considered more accessible were identified, and the students directed to select one of these:

- electricity – conductivity and resistivity
- fields and forces – the gravitational field strength at different distances from the Earth's surface
- quantum phenomena – demonstration of the photoelectric effect
- radioactivity – properties of alpha, beta and gamma radiation
- waves – diffraction of water waves and light waves

The delegates were allowed to work in pairs if they wished, and had access to support from the project's graduate assistants. These were Faculty students whom were either preparing for science teaching on the Post-Graduate Certificate in Education, or who were undertaking research degrees in science education. The graduate assistants' general brief for the project was to primarily act as observers, and not to seek to teach the delegates on the set tasks unless they specifically requested help. However, they were available to offer support whenever the delegates wanted to refer to them.

It was not expected that the delegates would be able to master all the ideas in one of the LO in the time available (about 90 minutes), but it was hoped that as able and motivated science students they would find the materials engaging and a useful taster for the types of material they would study if they selected physics as an A level subject after completing school science.

After each ASCEND session, the graduate assistants were asked to email in their observations of the session. The following comments are extracted from their reports on this session:

> "Around the room there were many absorbed expressions and good, interested questions were being asked. The program definitely encouraged thought and focus. ...The autonomy to find out and progress through for themselves in an accessible and presentable format gives the [gifted students] a huge selection of information and good opportunity to access it."
>
> "The atmosphere in the computer lab was amazing"
>
> "Everybody worked through the topics that you specified. So obviously they were captured by it, with no email-checking or laser-game playing ... I think the explanations were quite clear (certainly clear enough for those year 10s) and I only had a couple of really basic questions like, which symbol is gamma, and which is lambda."
>
> "All students were engaged with the program, and quite happy to work through it unaided ... they were able to navigate their way through it without any difficulty. I was surprised at how keen they were to get on with it. I noticed some did stop and think about what they were doing and talked to each other about it. Some students even questioned the validity of the physics they were presented with!"
>
> "I was impressed by their motivation, on the whole it was sustained for the full 90 mins."

The graduates also noted that they did not feel the software could have been used by these learners without some teacher support; that there was some skipping through sections (reflecting the experience with Adrian, above); that the mathematical demands were too high for some of the Y10 students; and that they suspected that some of the understanding was at a relatively superficial level. However, these points are to be expected given the apparent mismatch in maturity and knowledge between these children and the target audience for the materials.

The overall impression though was very much that the design Epic had adopted for the NLN materials had enabled these able younger learners to access and work with areas of physics beyond their school curriculum, and to do so with interest, enjoyment and enthusiasm.

CONCLUSION

This chapter has considered the design of educational software used to support learning in science. The chapter began with a consideration of the nature of ICT used in classrooms, and potential issues about the development, adoption and effective use of such technology in classrooms. It was suggested that there are barriers to effective use of ICT in teaching: including potential mismatch between what resources may offer, and what teachers feel can support them in meeting their educational aims and objectives. However it was also pointed out that, for example, in science ICT has strengths that offer potential to considerably enhance teaching and learning.

The main focus of the chapter has been on the use of ICT to provide learning resources for students (especially those that can be used with limited or no teacher supervision), and in particular on the need for these to be carefully designed in terms of pedagogy as well as subject knowledge. This is particularly so in science, in view of what is known about the challenges teachers face in encouraging conceptual change among students. Effective teachers work in a highly interactive mode in their classrooms, making on-line decisions based upon their subject knowledge, their pedagogical subject knowledge, their personal knowledge of the class and individual students; and taking into account contextual issues (how the pupils are today: irritable, tired, enthusiastic etc). Designing learning materials that could compete with teachers is currently a very unrealistic task.

However the chapter considered in some detail one example of learning materials written by experienced subject specialist teachers to a design template based on an instructional model. The Epic design template used for the UK's NLN Level 3 physics materials (commissioned to support teaching and learning in the FE sector), adopts many features used by teachers in planning their own work.

A case study of a Level 3 (A level) physics student using the materials for the first time was discussed in some detail. This raised some doubts about whether students using materials such as this in an inpendent learning mode would fully benefit from the design features. Despite this, the student clearly found the materials very useful, and with some 'teacher' input (probes and questions from the author) worked through much of the thinking process anticipated by the materials designers. His second experience of using these materials, some days later, was very positively evaluated – perhaps suggesting that familiarising with the design and format of such resources is important for effective interactivity.

Further, the use of the NLN materials with younger students who had not yet completed their basic school science, suggested that the design of the materials supported engagement and learning even with learners who had not yet mastered the previous stage of their education - showing that the intention that the materials could offer a useful 'taster' for students who may be interested in choosing physics at college levels (Epic, 2004) was effectively met. The students in this case were known (by virtue of being involved in the ASCEND project) to score highly on both interest in science, and science attainment – however, both these qualities would be expected of students admitted to A level physics courses in FE (the target users). More significantly, the successful engagement of these younger students with the learning materials took place in a context where they could work together and had teacher input available on-call. One of the features of this session noted by one of the graduate assistants was the level of conversation about the materials that was stimulated between students.

What this seems to suggest is that the challenge of producing learning resources that can be used for effective independent student study should not be underestimated. Students will 'miss' features, or deliberately 'skip' steps that look difficult, or where they know the answer is available at the touch of a button. However, as we saw with Adrian, learners may also quickly come to appreciate the design of materials that have pedagogic features carefully built into them, so some of this tendency to miss parts of the intended process could be overcome with strong induction into the use of resources. However, both in the case of Adrian, and in the ASCEND project, it was clear that the interactivity available in the software was more fully employed in the context of the learner having another human to refer to: whether a teacher or a peer.

The Epic designed UK NLN physics materials encompass an effective design, based upon sound pedagogic considerations, and clearly show that the design of educational software can be informed both by good subject knowledge and good educational thinking. The present work does however suggest that even well designed learning resources are likely to be most effectively used by dyads of pupils, or at least when students have some level of teacher supervision.

Perhaps in the future software development will progress further, or students' educational experiences will change (as they increasingly get used to working with ICT in school regularly from a young age), so that computer-based resources for fully independent learning start to offer a comparable experience to being taught by a good teacher. However the present state of the art should be reassuring for teachers: there are now some very well designed tools to support them in their work, but teachers are not yet in any real danger of being replaced by the ICT itself.

References:

Ausubel, D. P. (2000). *The Acquisition and Retention of Knowledge: a cognitive view*. Dordrecht: Kluwer Academic Publishers.

Becta. (2008). *Harnessing Technology - Next Generation Learning*, 2008-14. *A summary*. Coventry, England: British Educational Communications and Technology Agency.

Becta. (2009a). *Harnessing Technology for Next Generation Learning: Children, Schools and Families Implementation Plan 2009–2012* Coventry, England: British Educational Communications and Technology Agency.

Becta. (2009b). *Next Generation Learning - The implementation plan for 2009–2012: Technology strategy for further education, skills and regeneration.* Coventry, England: British Educational Communications and Technology Agency.

Branch, J. L. (2000). *The Trouble With Think Alouds: Generating Data Using Concurrent Verbal Protocols.* Paper presented at the Dimensions of a Global Information Science. from http://www.cais-acsi.ca/search.asp?year=2000

Bryce, T. & MacMillan, K. (2005). Encouraging conceptual change: the use of bridging analogies in the teaching of action-reaction forces and the 'at rest' condition in physics. *International Journal of Science Education, 27(6)*, 737-763.

Bugelski, B. R. (1971). *The Psychology of Learning Applied to Teaching* (2nd ed.). Indianapolis: Bobbs-Merrill Educational Publishing.

Chi, M. T. H. (1992). Conceptual change within and across ontological categories: examples from learning and discovery in science,. In R. N. Giere (Ed.), *Cognitive Models in Science* (Vol. XV, pp. 129-186). Minneapolis: University of Minnesota Press.

Davies, R. S., Sprague, C. R. & New, C. M. (2008). Integrating technology into a science classroom: an evaluation of inquiry-based technology integration. In D. W. Sunal, E. L. Wright & C. Sundberg (Eds.), *The Impact of the Laboratory and Technology on Learning and Teaching Science K-16* (pp. 202-237). Charlotte, North Carolina: Information Age Publishing.

Davis, E. A. & Varma, K. (2008). Supporting teachers in productive adaptation. In Y. Kali, M. C. Linn & J. E. Roseman (Eds.), *Designing Coherent Science Education: Implications for curriculum, instruction, and policy.* (94-122). New York: Teachers College Press.

diSessa, A. A. (1993). Towards an epistemology of physics. *Cognition and Instruction, 10*(2&3), 105-225.

Driver, R. & Erickson, G. (1983). Theories-in-action: some theoretical and empirical issues in the study of students' conceptual frameworks in science. *Studies in Science Education, 10*, 37-60.

Driver, R. & Oldham, V. (1986). A constructivist approach to curriculum development in science. *Studies in Science Education, 13*, 105-122.

Duit, R. (2007). *Bibliography - Students' and Teachers' Conceptions and Science Education.* from http://www.ipn.uni-kiel.de/aktuell/stcse/stcse.html

Epic. (2004). *Physics Level 3 Tutor Documentation v1.0.* Brighton, England: Epic Group PLC,.

García Franco, A. & Taber, K. S. (2006). *Learning processes and parallel conceptions – learning about the particulate nature of matter.* Paper presented at the 5th European symposium on conceptual change: Bridging the gap between mental models and situated cognition? Theoretical and methodological considerations.

Gilbert, J. K. & Watts, D. M. (1983). Concepts, misconceptions and alternative conceptions: changing perspectives in science education. *Studies in Science Education, 10*, 61-98.

Hennessy, S., Deaney, R., Ruthven, K. & Winterbottom, M. (2007). Pedagogical strategies for using the interactive whiteboard to foster learner participation in school science. *Learning, Media and Technology, 32(3)*, 283-301.

Heuer, D. & Blaschke, K. (2001). Learning physics with multimedia- and experimental-supported workshop instruction. In H. Behrendt, H. Dahncke, R. Duit, W. Gräber, M. Komorek, A. Kross & P. Reiska (Eds.), *Research in Science Education - Past, Present, and Future* (301-306). Dordrecht: Kluwer.

Karmiloff-Smith, A. (1996). *Beyond Modularity: A developmental perspective on cognitive science*. Cambridge, Massachusetts: MIT Press.

Kember, D. (1991). Instructional design for meaningful learning. *Instructional Science, 20*, 289-310.

Mercer, N., Dawes, Lyn, Wegerif, Rupert and Sams, Claire. (2004). Reasoning as a scientist: ways of helping children to use language to learn science. *British Educational Research Journal, 30(3)*, 359-377.

Miller, A. I. (1986). *Imagery in Scientific Thought*. Cambridge, Massachusetts: MIT Press.

Miller, G. A. (1968). The magical number seven, plus or minus two: some limits on our capacity for processing information. In *The Psychology of Communication: seven essays*, (21-50). Harmondsworth: Penguin.

Ofsted. (2009). *The importance of ICT: Information and communication technology in primary and secondary schools, 2005/2008*. London: The Office for Standards in Education, Children's Services and Skills.

Phang, F. A. (2009). *The Patterns of Physics Problem-Solving From the Perspective of Metacognition*. PhD Thesis. University of Cambridge, Cambridge. Available from http://people.pwf.cam.ac.uk/kst24/ResearchStudents/FatinAPBA.html

Songer, N. B. (2007). Digital resources versus cognitive tools: a discussion of learning science with technology. In S. K. Abbell & N. G. Lederman (Eds.), *Handbook of Research on Science Education* (471-491). Mahwah, New Jersey: Larwence Erlbaum Associates.

Taber, K. S. (1995). Development of Student Understanding: A Case Study of Stability and Lability in Cognitive Structure. *Research in Science & Technological Education, 13(1)*, 87-97.

Taber, K. S. (1998). An alternative conceptual framework from chemistry education. *International Journal of Science Education, 20(5)*, 597-608.

Taber, K. S. (2001). The mismatch between assumed prior knowledge and the learner's conceptions: a typology of learning impediments. *Educational Studies, 27(2)*, 159-171.

Taber, K. S. (2002). *Chemical misconceptions - prevention, diagnosis and cure: Theoretical background* (Vol. 1). London: Royal Society of Chemistry.

Taber, K. S. (2005). Learning quanta: barriers to stimulating transitions in student understanding of orbital ideas. *Science Education, 89(1)*, 94-116.

Taber, K. S. (2007a). *Enriching School Science for the Gifted Learner*. London: Gatsby Science Enhancement Programme.

Taber, K. S. (2007b). Science education for gifted learners? In K. S. Taber (Ed.), *Science Education for Gifted Learners* (1-14). London: Routledge.

Taber, K. S. (2008). Conceptual resources for learning science: Issues of transience and grain-size in cognition and cognitive structure. *International Journal of Science Education, 30(8)*, 1027-1053.

Taber, K. S. (2009a). Progressing Science Education: Constructing the scientific research programme into the contingent nature of learning science. Dordrecht: Springer.

Taber, K. S. (2009b). The Understanding Science Project from http://people.pwf.cam.ac.uk/kst24/projects/ECLIPSE/USP.html

Taber, K. S. & Riga, F. (2006). Lessons form the ASCEND project: able pupils' responses to an enrichment programme exploring the nature of science. *School Science Review, 87(321)*, 97-106.

Taber, K. S. & Riga, F. (2007). Working together to provide enrichment for able science learners. In K. S. Taber (Ed.), *Science Education for Gifted Learners* (182-196). London: Routledge.

Treagust, D. F. & Duit, R. D. (2008). Conceptual change: a discussion of theoretical, methodological and practical challenges for science education *Cultural Studies of Science Education.*

Warwick, P. & Kershner, R. (2008). Primary teachers' understanding of the interactive whiteboard as a tool for children's collaborative learning and knowledge-building *Learning, Media and Technology, 33(4)*, 269-287.

Watts, D. M. & Zylbersztajn, A. (1981). A survey of some children's ideas about force. *Physics Education, 16(6)*, 360-365.

White, B. Y. (1993). ThinkerTools: Causal models, conceptual change, and science education. *Cognition and Instruction, 10(1)*, 1-100.

Wood, D. (1988). *How Children Think and Learn: the social contexts of cognitive development*. Oxford: Blackwell.

In: Computer-Assisted Teaching: New Developments
Editors: B. A. Morris, G. M. Ferguson, pp. 63-92
ISBN: 978-1-60876-855-4

Chapter 3

INQUIRY WEB-BASED LEARNING TO ENHANCE KNOWLEDGE CONSTRUCTION IN SCIENCE: A STUDY IN SECONDARY EDUCATION

Manoli Pifarré Turmo*
Senior Lecturer in Educational Psychology at the University of Lleida (Spain)
Department of Psychology and Pedagogy, Faculty of Education.

ABSTRACT

Science is a complex topic and over the past two decades information and communication technologies have had an advantageous impact on science teaching and learning. When information and communication technologies are used into classrooms, students play an active role, learn at school and outside school, improve their teamwork skills, ask more questions, find answers to questions and show a higher interest in learning. The development of pedagogical strategies suited to classroom use of online resources should be an important priority for the science education community.

Although many studies have demonstrated the effectiveness of computer-based instruction in the students' achievement in science, others suggest that students get lost on the Internet when searching information. However, other works have reported that students have difficulties in regulating their learning when using hypermedia environments. Science educators claim the necessity of attempting to facilitate students' learning of science topics by using scaffolds, or instructional aids, designed to develop appropriate web searching and managerial strategies and to support students' web-based learning.

In response to this educational necessity, our work offers more insight into the challenge of combining information and communication technology with Inquiry Web-Based Learning and procedural guidance in order to increase the students' involvement and responsibility for their own learning of complex science topics. The main goal of the present study is to integrate technology and inquiry into science classrooms to assist students build their own knowledge and answer meaningful driving questions.

A challenging instructional approach was designed, implemented, and evaluated with 127 secondary education students. Specific scaffolds were provided to guide students

* Corresponding author: E-mail: pifarre@pip.udl.cat, (+34) 973 706570, FAX (+34) 973 706502.

through complex tasks, to help them develop scientific knowledge, and to provide support to enable them to transfer what they learnt during the web-based activity. All activities were designed keeping in mind the WebQuest structure but also including the circular notion of inquiry process presented by Lim (2004). According to Lim (2004) an inquiry process has the following elements: Ask, Plan, Know, Explore, Construct, and Reflect.

Quantitative assessments showed that students in the experimental group outperformed the ones in the control group in the post test, indicating the positive effect of the instructional design. Some clues to use adaptive scaffolds effectively were exposed. Some of these scaffolds are tables, simulators, applets, visualizations, conceptual maps, and note-taking areas/boxes.

This research lends evidence to questions regarding the value of students engaging in on-line inquiry learning to enhance science content understanding. The conclusions drawn in this work present a positive and promising path in the design of instructional processes capable to enhance students' digital literacy and, consequently improve student's content understanding using web information.

Keywords: Inquiry web-based learning, scaffolding, knowledge construction, secondary education.

1. INTRODUCTION

Educational research highlights the potential of the World Wide Web (WWW) as an instructional tool for classrooms. Since Internet can change the nature of learning it is important to stand out the next potentialities of the Internet: by increasing access to instructional materials in a variety of media, promoting students' skills in information gathering and problem solving, encouraging network collaborations, and having access to decentralized resources (e.g. Ryder and Graves, 1997; Hoffman, Krajcik & Soloway, 2003). The Internet offers education a window of opportunity to widen the traditional methods, materials and strategies of both learning and teaching. A survey conducted by Hsieh & Lin (1998) revealed that students are enthusiastic about using the Internet to help them in the classroom. They recognize its importance and usefulness as an educational tool, and are willing to face the challenges of cyberspace.

Science teachers and educators have the well-known aim of teaching science-related concepts meaningfully with the ultimate purpose of making students aware of how these concepts can be used in their daily lives. Science is a complex topic and, over the past two decades, information and communication technologies have had a favourable impact on science teaching and learning, as they have become tools to seek ways to make learning more meaningful, more transferable to different contexts, and more conducive to self-directed, life-long learning. According to Pelgrum (2001) when information and communication technologies are used in the classroom, students play an active role, learn at school and outside school, improve their teamwork skills, ask more questions, find answers to questions and show a high interest. According to Chang (2001) computer-based instruction plays an important role in contemporary teaching and learning of science concepts. There are several researchers who claim that student achievement in science increases with the use of computers (Chang, 2001; Coye & Stonebraker, 1994; Ferguson & Chapmen, 1993; Lee,

2001; Powell, Aeby & Carpenter-Aebyc, 2003; Rowe & Gregor, 1999; Tjaden & Martin, 1995; Tsai & Chou, 2002). Furthermore, Shute & Bonar (1986) argue that student abilities and skills in scientific investigations are affected positively by the use of computers in instruction.

Focusing on science learning, several researchers (Smith, 1996; McReary, Golde & Koeske, 2006; Shymansky, Hedger & Woodworht, 1990; Rubin, 1996; Prince & Felder, 2006) expound that inquiry learning has been frequently found to be more effective than traditional science instruction at improving academic performance, and the development of thinking, problem-solving, and laboratory skills. Lim (2004) says that scientific inquiry process is adopted to reveal the unusual natural phenomenon under study in science. In Inquiry Based Learning (–henceforth, IBL) students conduct science research as practicing scientists do. Since scientific investigation involves skills that must be learnt and subsequently practised, students are supposed to formulate hypotheses, develop their own designs, conduct self-directed experiments, collect data, assess their own progress, and analyze the findings in order to test the hypothesis and reflect on their inquiry process. This inquiry process is the most relevant factor which characterizes IBL and has been considered beneficial in augmenting meaningful learning. In the science classroom, IBL helps learners get involved with diverse ideas and knowledge integration and make connections between their existing ideas, information, observations, and diverse perspectives with the goal of developing more coherent and generative scientific knowledge (Martinello & Cook, 2000).

Science educators agree that the use of World Wide Web resources in classrooms as well as the use of IBL activities have a positive effect in learning science. However, the implementation of web resources to support IBL shows a few complexities that need to be considered. Some literature has highlighted that users often "get lost" too easily on the Internet when looking for information and that people have some difficulty when selecting the right information from the many sources of information available (Navarro-Prieto, Scaife & Rogers, 1999; Chui, 2002). Research also indicates that students experience certain difficulties in regulating their learning when they use hypermedia environments to learn about complex topics (Azevedo, Cromley & Seibert, 2004b; Brush & Saye, 2001). Wallace, Kupperman, Krackcik & Soloway (2000) conducted a study in middle-school science classrooms and observed that students found web technologies easy to use but they tended to reduce what researchers had envisaged as relatively open tasks, offering opportunities for personal engagement and intellectual challenge, to much simpler tasks of finding an obvious answer or a good website.

Researchers have therefore attempted to support the students' learning process by using scaffolds, or instructional aids, designed to promote the students' understanding (Palinscar & Brown, 1984; Paris, Wixson, & Palinscar, 1986; Rosenshine & Meister, 1992, Azevedo, Cromley & Seibert, 2004a). Scaffolds are tools, strategies, and guides designed to support learners in regulating their learning with hypermedia. Scaffolds can be provided by human and computer tutors, teachers, peers, and animated pedagogical agents during learning to enable learners to develop understanding beyond their immediate grasp (Chi *et al.* 2001, VanLehn *et al.* 2003). Several studies have provided evidence that learning about complex topics with hypermedia in the absence of scaffolds leads to failure to gain conceptual understanding (Greene & Land, 2000; Hill & Hannafin, 1999; Land & Greene, 2000). Thus, hypermedia can be used to enhance the learners' understanding of complex topics, like science, if they are provided with scaffolds designed to regulate their learning (Azevedo,

Cromley & Seibert, 2004a). Simons & Klein (2007) reported that students who used scaffolds performed significantly better than students who did not and their answers were, by comparison, more highly organized. Studies have demonstrated that scaffolding in biology, statistics, and science leads to enhanced learning for middle school and college students with hypermedia (Kao & Lehman, 1997; Pedersen & Liu, 2002a, b; Azevedo, Cromley & Seibert, 2004a). Thus, these results suggested that scaffolds may enhance inquiry and performance, especially when students are required to access and use them.

Overall, although most studies have demonstrated effectiveness of computer-based instruction in students' achievement in science, others suggest that students get lost on the Internet when searching information. Studies have also reported the effectiveness of inquiry learning at improving academic performance and meaningful learning. However, other works have reported that students have difficulties in regulating their learning when using hypermedia environments. There is no doubt about the fact that well-designed IBL activities on the web provide learners with cognitive tools and procedural guidance for their inquiry and help them form a learning community wherein teachers and learners interact to solve complex problems (Slotta, 2002). Science educators claim the necessity of attempting to facilitate students' learning of science topics by using scaffolds, or instructional aids, designed to develop appropriate web searching and managerial strategies and to support students' understanding.

In response to this educational necessity, our work offers more insight into the challenge of combining information and communication technology with IBL and procedural guidance in order to increase the students' involvement and responsibility for their own learning of complex science topics. The study described in this paper is aimed at helping the science education community move toward extending the notion of designing and implementing scaffolding in inquiry web-based activities. With this in mind, an instructional approach to help students understand web information and learn science contents from the web was designed, implemented and evaluated during one academic year with secondary education students. On-line learning materials focused on specific curricular science contents and with specific scaffolds which help students accomplish web-based tasks and construct their knowledge were designed by the authors. The effect of the students' participation in the IBL instructional process was analysed in terms of the students' learning processes and outcomes.

Section 2 of this chapter summarizes the main instructional variables that have guided the design of our web-based activities: Scientific Inquiry Web Learning and the design of a scaffolding process to guide students' science knowledge construction. This section is illustrated with examples of the different web-based activities. Section 3 explains the entire research process that enabled to analyze the impact on the learning outcome of students using the web-based activities designed for this study. Finally, section 4 shows the main conclusions of the present study.

2. Inquiry Web-Based Activities: Main Instructional Variables to Design the Digital Activities

With the purpose of promoting students' skills in managerial web information to learn complex science topics in a meaningful way, web-based instructional materials were designed by the research group. Scientific Inquiry learning, scaffolding, and knowledge construction were key concepts that guided the activities' design process. In the following sections we explain how we promoted these key concepts in our activities and how we designed the web-based activities.

2.1. Scientific Inquiry Web Learning: WebQuest Methodology

From our point of view, authentic knowledge involves participating in real science, which is similar to the science that actual scientists perform. Our design process focused on building and reinforcing scientific skills and knowledge construction in order to teach students to think systematically about the process of scientific inquiry.

Scientific inquiry was implemented in the activities by following this progression: (1) asking research questions: students had to face interesting questions connected with their previous knowledge and real life, (2) students were asked to set out their hypotheses, (3) the experimentation was always web-based and students were asked to plan for their work and collect data, (4) students compared their initial hypotheses against their actual results. Students were asked to figure out whether their predictions were correct or incorrect and state why. They were invited to revisit questions and criteria, as well as clarify their specifications, (5) finally students applied or transferred their learning to different contexts, elaborating a Final Product that would wrap up the activity.

Inquiry Based Learning is a teaching strategy that expounds and solves real-life problems. The teacher's task consists in selecting problem situations and guiding the students into finding the most comprehensive and significant conclusion available. The students need to understand the problem and its consequences in order to take the necessary steps towards its final solution. To achieve this end, the problem must be set out in such a challenging way that students may be keen to pursue and, at the same time, make them feel that they are able to solve it. This teaching method allows students to find an easy link between scientific knowledge and real life. The standard question *what do we need to study and learn this for?* finds in this new learning strategy a new answer available to the educator.

In our work, the problems set out imply using, analysing and understanding web-based information. Our work introduces the internet as an element that may contribute to developing specific skills and cognitive processes. Our work intended to show the use of technology and the web as a support to the students' mental activity. Several authors highlight the educational use of the web as a tool that increases the possibilities of IBL (Martinello & Cook, 2000; Pelgrum, 2001; Lim, 2004). Among the key aspects that using the web may promote and may contribute to more constructive and meaningful learning, the following are worth noting:

- The web may increase the possibility of active participation of students. For example, the students have a large amount of information available and several ways of processing it.
- The use of web information may broaden self-regulated teaching in that the students decide on the aspects of information they choose to either analyse in depth or study. Presenting the information in a hypermedia or hypertext format allows the student to

decide on the paths he wants to pursue and the information he needs to analyse in depth.

- Likewise, more participation of the student in the learning process and more decision and self-regulation of this process facilitates that he starts up a series of key cognitive processes and strategies to solve intricate problems
- Finally, using the web enables further collaboration and communication between students and people that may be relevant in learning specific contents.

With a view to boosting the above features offered by web-based work, our project has used and adapted WebQuest methodology. The WebQuest work methodology helps promote Scientific Inquiry Based learning as this teaching/learning methodology is based on presenting and solving real and significant problems to the student. Solving these problems implies an active role from the students and the starting up of mental strategies of a high cognitive level that allows to organise the information, analyse it and come up with new answers.

This teaching methodology uses the computer to integrate constructivist learning principles and follows teaching/learning tenets based on problem-solving as well as uses web resources to develop the curriculum. WebQuest implies investigation and research on the web. WebQuest teaching methodology was originally set out in the middle of the 90s by Bernie Dodge (1995). The initial philosophy was to develop in the students the capacity to explore the internet with a clear aim in mind: learning to select and retrieve data from multiple sources and develop critical thinking skills (Dodge, 1997). Essentially, a WebQuest consists in presenting the students with a problem together with a series of resources pre-established by its creator. The educator tends to be the one in charge of guiding the student by preventing from aimless navigation on the web.

The main basic features of a WebQuest activity are the following: it is an extremely simple and enriching teaching model that fosters the educational use of the Internet through cooperative learning and research processes to learn. It presents a research activity whose information is mostly downloaded from the internet. This research mostly ends with the elaboration of the final product, namely a web page, a power point or text document with the results of the research. It fosters learning based mainly on Internet resources and encourages students to investigate and develop critical thinking, creativity and decision-taking. In general, it may contribute to develop different learning skills by helping students transform the knowledge acquired.

Taking into consideration the approaches set out in our study, the design of a WebQuest activity and its implementation in class has been guided with the following four educational principles: a) structuring and organising the students' work; b) motivating and authentic learning; c) cognitive development and scaffolding and d) collaborative learning. These principles are developed as follows:

a) Structuring and organising the students' work to solve the problem set out

A WebQuest activity sets out an intricate problem that students need to solve and this requires the starting up of an equally complex and intricate solving process. In order to facilitate this process and solve the problem successfully, WebQuest design emphasizes the need to structure and organise the students' work. Although there is no fixed format for the design of WebQuest activities, which is one of its main advantages, most WebQuest activites contain a series of stable features that help organise the teaching/learning web-based process. The main features of a WebQuest that we have taken into consideration when creating our activites are shown in figure 1 and are the following

Firstly, the **introduction** presents the challenge or the problem together with the background context. This section aims at getting students to activate their previous knowledge on the topic at hand in order for them to be able to develop that activity and make them ready for that new learning methodology. An example of the introduction section presented in our web activities is exposed in figure 2.

Next, the **task** or tasks are set out. This is one of the main parts of the WebQuest as it rounds up the learning aims of that activity. The **process** indicates the steps to follow to achieve the aim set out in the activity, namely solving the problem/challenge or carrying out the task at hand. It is at this point that students are given clear details on what they have to do, as well as when and how to do it. This process includes the **resources** that the students may need both during the stages of receiving information and transforming and elaborating the final product. The **assessment** is the stage that specifies the evaluation criteria of the activity. It is done by means of patterns whereby the student is always informed on how his work will be assessed. This helps guide and motivate the students during the activity. The assessment pattern may be applied to the whole WebQuest or to each individual section. This pattern may be applied to each student individually or to each group. The last stage of the activity contains the **conclusions**. On this stage, the initial project is analysed again and the students are invited to reflect on what they have learnt during the WebQuest activity. This stage also specifies the final destination of their work (e.g. uploading it on the internet or sending it to the relevant authorities) and presents practical examples of how the students may explore further on the topic and apply the newly-learnt strategies and knowledge to other areas.

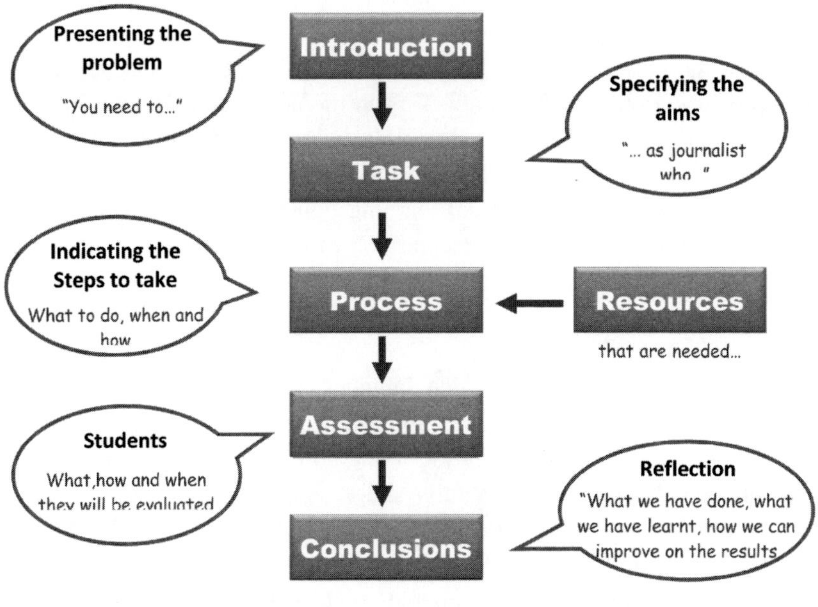

Figure 1. Main didactic elements of a WebQuest.

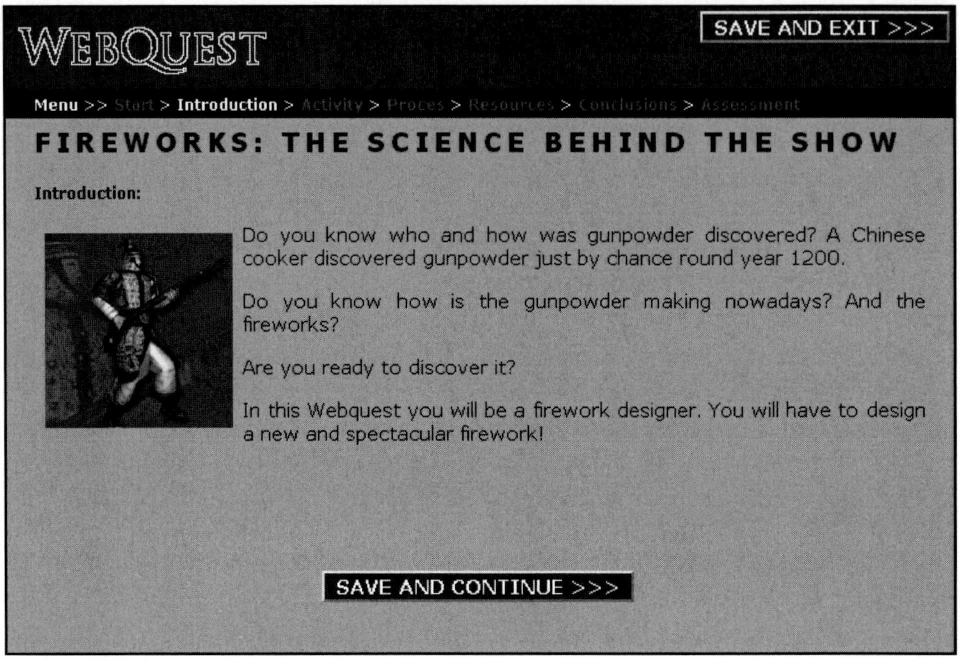

Figure 2. Example of Introduction section of the web-based activities.

b) Motivating and authentic learning

Another educational element of WebQuest activities is promoting authentic learning. WebQuest activities are based on one actual question whose answer must be supplied by the student. To come up with the right answer, the student will use the information obtained from a series of internet resources or other resources that may have an impact on society. The student will also use the options offered by the internet and publish or send them to other people and, subsequently, obtain some feedback from them. The authenticity inherent to the design of an activity using WebQuest methodology may help motivate the student as it displays a problem that he can relate to and create a bond with his previous knowledge on the topic. All these contribute to create an ideal learning context for the development of high cognitive learning processes that may be crucial in promoting learning that is significant, autonomous and applicable to other situations.

c) Cognitive development and scaffolding

Solving a proper WebQuest, i.e. one that includes solving an authentic and intricate problem, means that the students need to start up high cognitive processes (or high thinking levels) in order to organise information, analyse it and come up with new answers. To solve the problem set out in a WebQuest successfully and to ensure that the student will develop higher cognitive processes, he will be provided with a series of scaffolds in the form of guidelines or instructions that will make it possible for students to carry out the task at hand. Developing a series of adequate and suitable scaffolds that properly sequence the teaching/learning process is one of the main challenges set out in the design of our web-based activities. They are accounted for in section 2.2. of this chapter.

d) Collaborative learning

This type of learning is based on one member of the work group achieving his aims as long as the rest of the group achieves theirs. This does not imply a simple accumulation of interventions but rather group interaction with features like positive interdependency, individual responsibility, stimulating face-to-face responsibility, group assessment, simultaneous interaction and similar possibilities of participation from the students. This type of learning implies that all members are necessary to the group, which reinforces the students' self-esteem. Generally, a WebQuest is not designed to be solved individually. The intricacy of the problem and the sources that must be looked up recommend collaborative work between several students. Collaborative learning is a key element in designing WebQuest activities. They sometimes set out activites where each student is responsible for one part of the solving process or a specific role. On other occasions, they set out activities whose solution requires the contribution from the rest of the students in order to achieve the final aim

2.2 Scaffolding: key Feature in the Process of Knowledge Construction

The classic design of WebQuests requires features that may help solve the problem set out such as organising the task into sections and facilitating web-based resources to solve the activity. However, the activities are not designed with explicit help tailored to the activity at hand. In our view and, taking into consideration constructivist principles, we understand that we need to incorporate those devices that may have a positive educational influence on the process of knowledge construction from the student. The key issue is, *how can one teach what needs to be constructed?*

Our study presents the design of a series of scaffolds that guide and enhance the students' learning processes. Our project is based on pedagogical scaffold as a positive educational mechanism that is essential to achieve a significant process of meaning construction that is close to specific cultural knowledge (Palinscar & Brown, 1984). Pedagogical scaffold acts as a mediator between the students' mental activity and the knowledge they want to acquire Consequently, when creating the WebQuest activities of our study, we have designed and sequenced a series of scaffolds referring to different learning competences and processes needed to solve the problems set out at the beginning. Designing, sequencing and presenting the pedagogical scaffolds for the activities take into consideration the constructivist perspective of teaching/learning processes (Palinscar & Brown, 1984; Paris, Wixson, & Palinscar, 1986; Rosenshine & Meister, 1992) and are summarised as follows

a) Scaffolding should only be an assistance but never a substitute for the student's mental activity in the process of meaning construction and sense assignation
b) Following the tenets of knowledge construction as a process, scaffolding must be conceived as a process whose help is suitable in quality and quantity to the student's needs. The dynamic quality of adjusting scaffolding is of paramount importance. Therefore, efficient educational intervention implies adjusting the help by starting at the level that the student departs from and setting out achievable challenges that move beyond his initial level.
c) To ensure a scaffolding better adapted to the needs of the different students, it is necessary to diversify the type of help offered

d) Scaffold must integrate the learner in the process from the initial stages of the activity
e) caffolding must be temporary and, therefore, must be withdrawn gradually as the learner starts taking responsibility for his work.

Considering these five basic criteria, the type of scaffolds offered to students is presented below together with an explanation of how these activities are presented and how this help is gradually withdrawn.

2.2.1. Types of scaffolds

In our activities, the students need to search and manage digital information under two situations:

1. Some situations pre-select a series of web sites, specific webs loaded with relevant information. The students have to surf, search and select relevant information that enables them to solve an activity using a limited range of web resources.
2. Other situations fail to offer students web resources to solve the problem at hand. In this case, the students will have to use search engines on the web and surf, search and select from a wider range of web resources to obtain relevant information and solve successfully the task at hand. In the latter situation, the students have to take a higher number of decisions on which information to search, how to define a better search strategy, how to refine a search strategy according to the results obtained and which criteria to use to select information that is relevant to solving the task. As this is a difficult process, we have introduced a series of psychopedagogical digital scaffolds that may guide the students. How did we guide the students' work process? Generally, the guidelines designed consist in a series of questions whose answer implies that the students have previously set out and solved key aspects of the search, selection and treatment of web information that will enable them to solve the problem at hand.

What processes and skills are key to solving problems that use web-based information? One of the main aims of our instructional process is that students may increase and optimise the efficiency of their search, select and organise competences of web-based information. To this end, a series of helping devices were designed -scaffolds. All the activities were fitted with guidance on the following five skills that may help solve digital information, which we understand are basic and essential to solve successfully the activities set out (Eisenberg & Berkowitz, 1990; Brand-Gruwel, Woperis & Vermetten, 2005; Pifarré, Sanuy, Vendrell & Godia, 2008). The processes are the following:

1. Searching for information
2. Selecting the digital information found
3. Organising individually and significantly the digital information needed to solve the problem.
4. Using the digital information individually to solve the problem.
5. Regulating and adjusting the solving process.

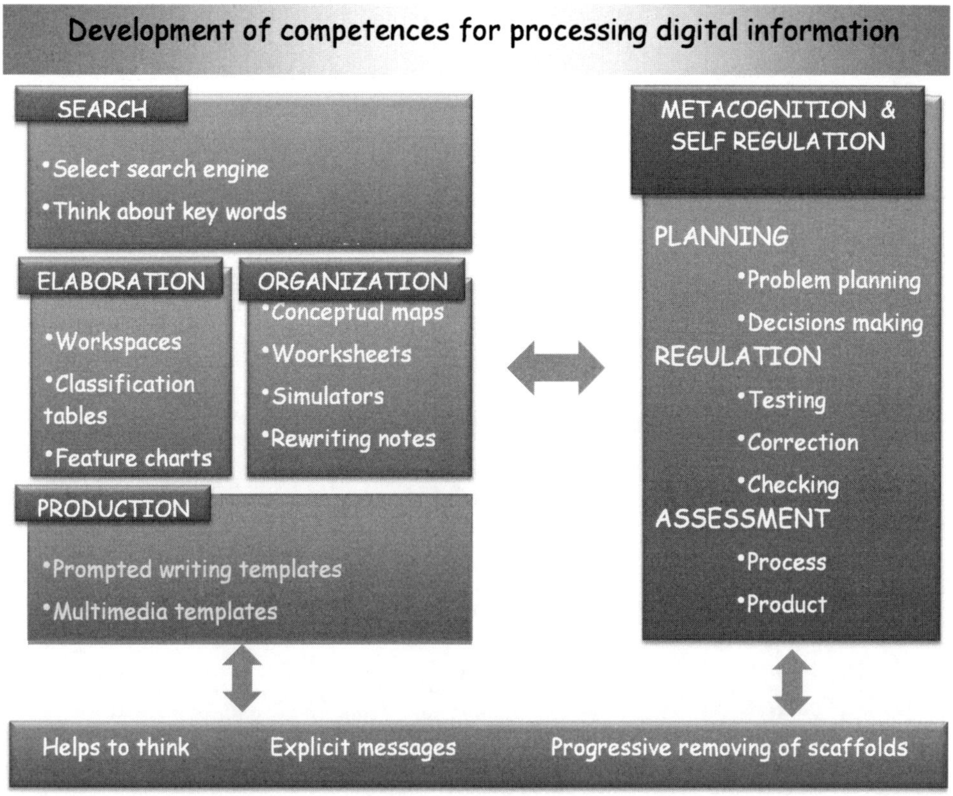

Figure 3. Digital competences to process digital information.

Figure 3 shows the diagram of the digital competences designed. Details and examples of how these competences have been developed in our web-based activities are explained bellow.

Scaffolding the process of searching for information

The scaffolds designed to develop competences related to the process of searching for information revolves around the following five competences:

- Defining the problem. This type of help is designed to get students to specify the need to search for specific information and activate previous knowledge on the topic that will help define the processes following the selection of a search engine, the key words and the information source.
- Choosing the best search engine.
- Selecting the most appropriate key words to search information on a specific topic.
- Efficient use of the advanced search options offered by the different search engines, for example, language selection and use of Boolean searches.
- Refining the search strategy

An example of scaffold related to defining the problem and searching information is presented in Figure 4. The example is an extract from the activity called "Management of toxic waste" (Pifarré *et al.* 2007) . In the first module of this activity, the students have to search information on the safety regulations required for the transport of toxic waste. The students are asked a series of questions that they are required to reflect on and specify how they will search for this information on the web. To be precise, the activity asks students on four variables in the process of information searching What they have to search, i.e. what kind of information they need to search. What search engine they will use and why. What type of web pages they will find and their characteristics as well as the criteria they will use to assess their suitability and solve the problem.

What web pages are more appropriate to solve the activity. In other words, what web pages they will select to read and search for the information specific to that activity

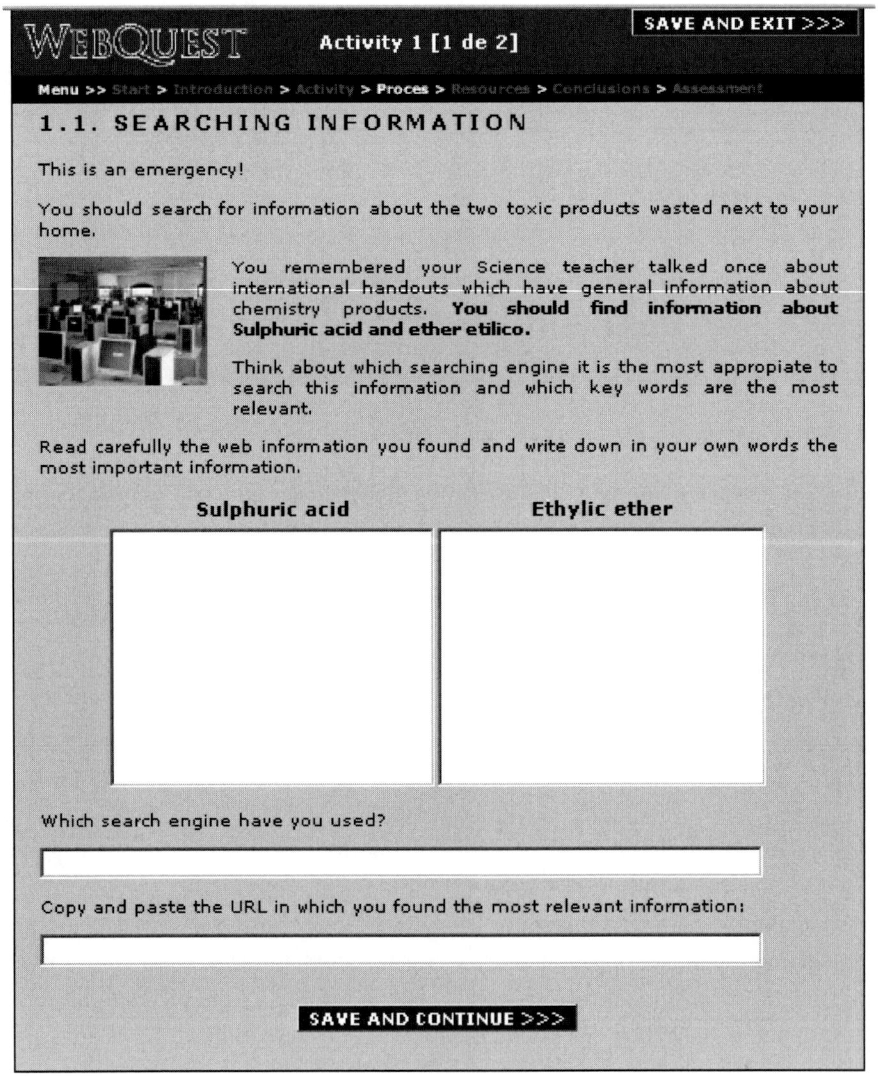

Figure 4. Example of scaffolds to develop search competences of digital information.

Scaffolding the process of selecting and elaborating digital information

Once the student has found the most suitable web page, he needs to understand and select the information most relevant to answer the question set out in the activity. Therefore the student needs to develop strategies that may help select and elaborate this information in a way that is personal and adjusted to the task and problem set out in the activity. In order to guide the student, our digital activities have developed scaffolds that enables students to personalise the information obtained. Personal elaboration of information implies integrating the information materials by relating the new information to the information previously stored in memory.

The following scaffolds would be suitable at this point:

- Scaffolds that enable students to analyse the conceptual structure of information of a web site. For example, formulating a series of questions related to the features of the content lists and indexes of a web page
- Specific area/boxes – text tables to take notes on the relevant information of a web page. The students have to indicate the relevant contents and account for their choice
- An area/box to generate ideas. After reading and selecting relevant web information, the students are asked to generate ideas or solutions to the problem. This process may be called "brainstorming".
- Tables or diagrams are made available on the main topics of the information that has to be searched. The students need to complete these diagrams with the information obtained in the web. These tables and diagrams are geared to help students classify their ideas and handle more easily the information obtained.

Figure 5 illustrates the area designed to take notes from the information obtained in the web. To be specific, this example is an extract from the activity called "Experimenting safely in the lab". The students have to extract and take notes of a series of official web pages that offer information on the main safety regulations to observe when working in science labs.

Scaffolding the process of organising digital information

Once the information has been individually selected and elaborated by the student, the next step consists in organising all the material gathered with a view to answering or solving the problem. Therefore, the ultimate objective of organising digital information is combining the information elements selected in a coherent, significant and individual way.

The main scaffolds available to the students in order to organise the information is the following:

- Conceptual maps. They are used as strategies that organise and represent knowledge. Besides, conceptual maps help establish relationships between the different concepts related to a topic and help find the way to solve problems.
- Work sheets (spreadsheets). Many activities in the field of maths and science have used spreadsheets as a tool to help students on three aspects: a) organising numeric information according to specific variables; b) contrasting initial predictions against real data and c) representing numeric data into graphics in order to account for them more personally

- Simulations. Simulations call for high cognitive requirements as they represent study situations with intricate real situations. In our activities, the use and analysis of simulations is guided. Students are asked to answer and reflect on aspects related to planning, the specific use of steps and actions to use the features of a specific simulation effectively and their results. These steps consist in: giving an answer to specific questions by means of simulation, guiding the attention of students on specific aspects of simulation, searching and using previous information as a tool to interpret what is happening in the simulation.

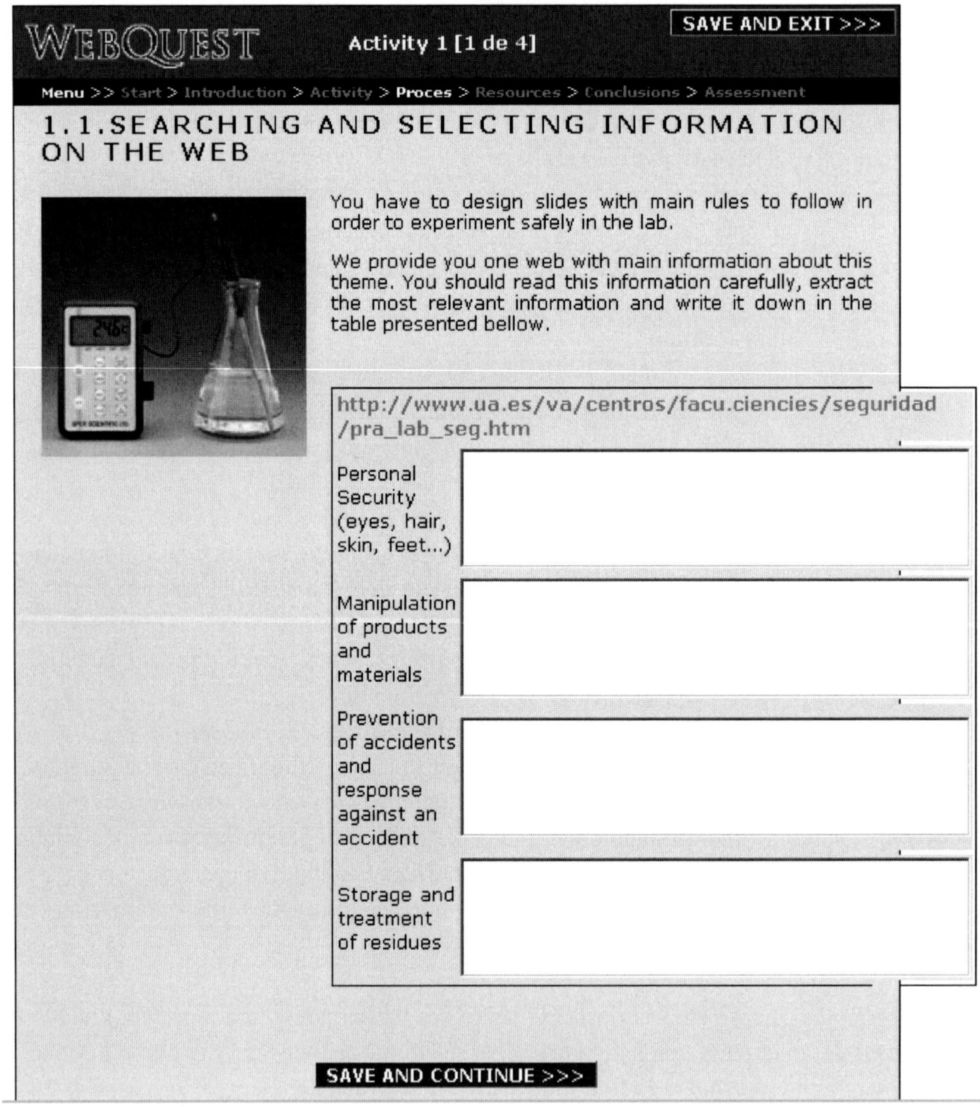

Figure 5. Example of scaffolds for the development of selection and elaboration competences of digital information.

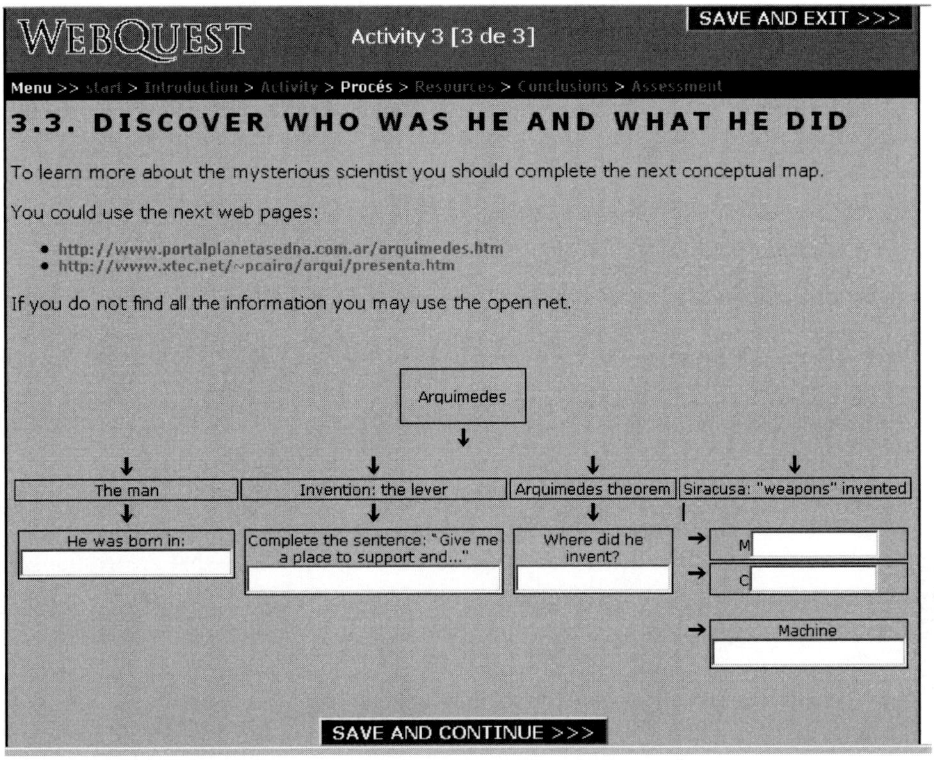

Figure 6. Example of help for the development of competences of digital information organisation.

For secondary school students, organising the research data obtained during the activity might be a difficult task. This is why they are offered models on how to organise the data significantly. This scaffolds requires from students to complete flowcharts, charts, diagrams and conceptual maps. Figure 6 illustrates a conceptual map which requires from students to search and organise information on the discoveries of the Greek scientist Archimedes and on the principles that regulate the laws of flotation of solids and liquids.

Scaffolding the process of using digital information to produce the final product of the activity

All the activities aim to solve a problem. The solution must be specified through a specific learning object or a final product in the solution process of the activity. This final product represents the students' knowledge object or product of the learning activity. There are several resources that allow to elaborate a knowledge object or product of the learning activity, as in our digital activities the students must obtain the following formats as the final product.

- Writings. The students work on several types of writing documents, e.g. manifestos, newspaper articles and letters…
- Multimedia presentations
- Tables of contents
- Multimedia guide sheets
- Solving mathematical problems

To elaborate the final product, the student obtains scaffolds to assist him in the essential aspects of the activity which provide an answer to the problem at hand. This scaffold guides the thinking process and the various steps to follow to reach the final product. Depending on the format of the final product, the students are provided with a template that helps them plan and organise the content of the final product. Depending on the type of writing document – letter, newspaper article, manifesto- and the the presentation format–powerpoint, mural, poster, ... each activity give to the students specific scaffolds in order to construct the final product.

Scaffolding the process of regulating the overall digital activity resolution

Several education forums have highlighted the importance of the regulating and control processes of one's knowledge and/or one's cognitive process of learning. Several educational proposals have been developed that encourage students to manage and regulate efficiently those actions that allow them to solve successfully a specific task. To accomplish this ambitious aim, it is important to highlight the need to not only boost processes to plan, control and assess the learning activity but also develop processes that are responsible for regulating and controlling the cognitive processes.

Generally speaking, planning is the organisation previous to the proceses and actions that will make the task easier. This process implies designing and selecting the most appropriate procedures to solve a task and allows to structure and organise the solution. At this point, two planning subprocesses can be found. On the one hand, the planning process or metaplan, also known as core strategy. This subprocess constitutes the most general metacognitive strategy related to the series of decisions that allow facing the general problem. On the other hand, the second subprocess is decision-taking on the most appropriate actions and their execution sequencing to achieve the task aims.

The control processes show up when solving the task and may refer to verifying, rectifying and revising activities of the strategy used.

The assessment processes are those results referring to actions taken once the task is solved and aim to contrast the results of the strategy used in terms of efficiency. When designing our digital activities, a series of pedagogical scaffolds devices were implemented which aim to get students to solve the activities meaningfully and by guiding them through their cognitive resources. To this end, the following aspects need to be taken into account:

a) The solving process must be divided into the following stages. Generally speaking, these stages correspond to three phases: detailed planning of solving the activity, regulating and controlling its solution, and ensuing assessment.
b) During the solving process, students are explicitly asked on the different relevant variables of the context of the task solution. The students need to express their opinion on how they take all these variables into account in the solving process. The students are asked questions on the variables to take into account in the decision-taking process that allows solving complicated problems set out in the activity. Explicit messages are introduced that alert and guide the students through key actions of the task solving process.

2.2.2. *Presentation of scaffolds*

The scaffolds suggested to students is designed to guide and enrich the solving process and aim to develop the different digital competences presented in the digital activities by using three methodological strategies: thought-provoking sheets, introducing explicit messages and planning for the gradual withdrawal of help. The three strategies are explained as follows:

a. Thought-provoking sheets

The digital activities set out questions, indications and suggestions that force students to think on the actions and variables required to solve the task. The questions that the student needs to ask himself and give an answer to are designed to foster explicit work on the different procedures and strategies followed in order to come up with a solution. Generally speaking, the questions asked to students aim to:

- Make them familiar of and aware of the necessary procedures to approach the solution to the problem of the digital activity.
- Make them apply these procedures when solving the task.
- Make them analyse the quality of the task solving and reflect on the different conditions that require the use of a specific solving procedure, the implications of the solving process and the final product. The ultimate aim of this reflection is to make learning transferable and functional in the solving process of problems and tasks.

b) Introducing explicit messages at key and relevant times of the solving process.

At key times of the solving process, the students are presented with a message that refers to a specific learning procedures that is needed to solve the activity successfully. This message comes up in a different format from that of the rest of the information of the activity, i.e. it clearly stands out and is shown as a brief message in a speech bubble.

We have designed as many as 6 groups of explicit messages for students. Each group refers to cognitive procedures related to searching, processing and using web information or the regulating procedures of the solving process.

c) Gradual withdrawal of scaffolds

As pointed out above, pedagogical scaffold may be a key element that boosts knowledge construction of students. To achieve this, special care must be devoted when designing the type of scaffold provided and how it will be provided. Furthermore, another requirement of pedagogical scaffold is its temporary nature. In other words, activities must be designed with a view to eventually withdrawing them in order for the student to be able to solve these tasks individually and without any assistance. This is why, in our study, both the amount of help and its typology vary in time. The first activities contain more help messages and are longer. As the learning progresses, the help messages diminish and the student is endowed with more responsibility in the solving process

3. Research Study: Analysing the Impact of Inquiry Web-Based Learning In Science Knowledge Construction

Recent psychology research has started to investigate the cognitive load required to take advantage of digital information for learning purposes. This research demonstrates that access to digital information requires the development of specific and complex cognitive processes. Learning with digital resource-based environments depends upon the students' mindful application of a variety of learning strategies, including selecting, organising and integrating information with prior knowledge. One of the characteristics of digital information is the hypertextuality which allows nonsequential reading and thinking. Many researchers have studied the cognitive needs that allow using effectively the information presented in hypertext format (De Jong & Van der Hulst, 2002; Rouet, Vidal-Abarca, Bert-Erboul & Millogo, 2001) and they have shown that the effective use of a well-structured table of contents increased the students' ability to find relevant information in a science task

However, many studies also indicate that students are ill-equipped to search, use and integrate digital information (Lazonder, 2000; Guinee, Eagleton & Hall, 2003; Lazar, 2003; Tabatabai, 2004). In this context, an important step in helping students to use the Internet more effectively as a learning tool is to understand their baseline searching strategies. In a previous work (Pifarré, Gòdia & Martorell, 2006) we investigated strategies used by teenagers to search, comprehend and integrate web information in order to solve learning tasks. Sixty-eight 2nd year students of secondary education (13-14 years old) participated in that study. All of them solved individually a curricular activity using the World Wide Web.

Results showed, firstly, low students' success in solving the curricular tasks. Secondly, the students did not take full advantage of the search engines' potential, e.g. they ignored the "refine" and "broaden" commands; they failed to use Boolean operators or select right key words. The students' search for information on the web was data-driven. Thirdly, our study revealed that students did not develop planning strategies to search for information on the web. Finally, students did not integrate the web information found on the Internet and the most common strategy used by students to answer the curricular questions proposed in the task was "copy and paste".

In this context, the necessity to generate instructional environments to integrate the World Wide Web as a learning tool in the curriculum and enhance the metacognitive domain and system knowledge became a critical issue to improve the students' web performance (Land & Greene, 2000). As a response to this feedback, the main goals of our research study revolved around designing, implementing and evaluating an instructional approach that would help students focus on the content developed on on-line learning materials to provide scaffolding. This allowed students to accomplish web-based tasks and to support the students' information-seeking activities as they asked meaningful questions, searched for information, organised and assessed their findings, as well as created rich representations of their newly constructed understandings.

We designed a year-long study to investigate the depth and accuracy of secondary students' knowledge construction in Science as they utilized on-line resources to solve instructional tasks.

3.1. Research Questions

This work focused on the following research questions:

1. Are secondary education students successful in finding and integrating web information to construct knowledge in complex science topics at the beginning of the study?
2. Are secondary education students successful in finding and integrating web information to construct knowledge in complex science topics after their participation in the web-based instructional process?
3. What is the effect of the features of the instructional task on the students' success in finding and using web information to learn science?

3.2. Method

3.2.1. Participants

One hundred and twenty-seven (n=127) 2nd year (13-14 years old) and 4th year (15-16 years old) students of secondary education from three different schools from a middle class urban environment in Lleida (Spain) participated in the study during one academic year (2005-2006). Students had little knowledge of the science-related topics covered in this work. Details of the participants and the experimental design are shown in Table 1.

Two institutions were randomly assigned to the experimental group, where students followed the web-based instructional process during one academic year. The third institution was taken as the control group and its students continued their instructions using the traditional teaching method.

3.2.2. Procedure

The experimental process consisted of three phases: pre-test, intervention and post-test as presented in figure 7

Table 1. Participants and experimental design.

Institution	Group	Year	n	Pre-test (Mars)	Intervention: Web-based activities	Post-test (Moon)
Institution 1	Control	2nd	24	√	X	√
		4th	15			
Total Control			**39**			
Institution 2	Experimental	2nd	21	√	√	√
		4th	20			
Institution 3	Experimental	2nd	23			
		4th	24			
Total Experimental			**88**			

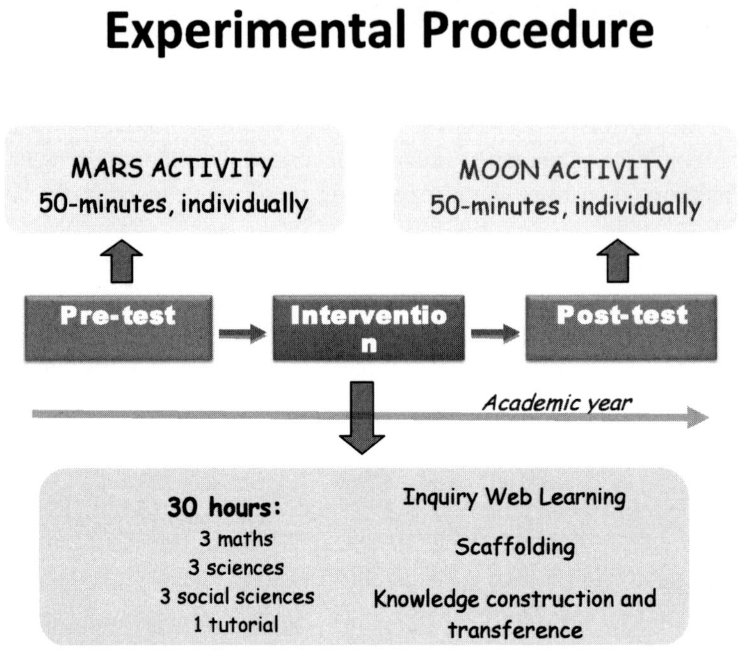

Figure 7: Experimental procedure followed in the research study.

Pre- test: the mars activity

A pre-test web-based activity focusing on planet Mars was given to all students and solved individually in a fifty-minute session. The content of the activity was new to all the students. The students' answers were stored in a server, scored and analyzed by the author to evaluate the secondary education students' performance in the process of finding and using web information to solve curricular tasks and to study their depth and accuracy of understanding. The pre-test was organized in two different tasks:

Task 1: Fact-finding task. In this task students had to search for specific data on the web about planet Mars such as temperature, gravity or number and names of satellites. Students had a conceptual map with 14 questions to answer. To solve the task some web resources were provided. Students could either use these resources or the open net.

Task 2: No-Fact-finding task. This task was divided into two sections: a) students had to find information to answer three general questions about planet Mars: its climate, the main exploration missions in the past and the main findings of these missions. Students in this task were expected to find the right textual information, read it carefully, and select the most appropriate and synthesize it. To this end, some web resources were provided; b) In section 2, students were invited to answer the question "Is it possible to live on Mars". No web resources were provided for this question and students had to use web browsers, develop specific searching strategies, select the right information, and write out a final text accounting for their answer.

Table 2. Science content of the web-based activities.

Season	2nd year	4th year
Fall	Atmospheric Pressure	Waste Management
Winter	Laboratory Safety	Chemical Reaction
Spring	Buoyancy (Archimedes Principle)	Electromagnetism

Intervention

The experimental group was engaged in a year-long web-based instructional project consisting of a total of 10 web-based activities in three areas: science (3 activities), maths (3 activities), and social science (3 activities) plus a tutorial searching activity. These activities were solved in pairs in three sessions of fifty minutes each. For this study, only science activities were considered.

Six web-based activities from the 2nd and 4th year science curriculum were designed. They are presented in Table 2.

Each web-based activity consisted of three tasks that guided the students to give a final answer to the essential question of the activity. *Fact-Finding* and No-*Fact-Finding* tasks were introduced in all activities.

The subtasks included *Fact-Finding* and *No-Fact-Finding* activities ranging from the easiest to the most difficult. Students had to through each of the three subtasks, completing activities that allowed them to develop the background knowledge required to search, collect, analyze, interpret, and present their data effectively.

Fact-Finding tasks were simple activities which required from students to search for specific data on the web. Specific data refers here as temperature values, a specific date, number or name of specific items. To solve the activities, students were provided either with web resources or the open net. On the other hand, *No-Fact-Finding* tasks were more complex. Students were asked to answer more general and open questions and were expected to search for that information on the web, find the right textual information, read it carefully, select the most appropriate one and synthesize it.

The sessions taken to solve the activity were divided into three phases. An opening phase (5-10 minutes) where the teacher welcomed and settled the class as its members arrived and introduced the task to be undertaken. In the seatwork phase, students worked on the tasks assigned and were assisted by the teacher. In the closing phase (less than 5 minutes) the lesson came to an end and the students' work was stored in the server.

As presented in section 2 of this chapter, the design of all the web-based activities was intended to support question asking, develop efficient searching and managerial strategies and help the construction of new understanding.

Post-Test: the Moon Activity

A specific computer-based activity was designed in order to study whether the search and managerial strategies used by students in the process of finding and using web information had improved. As in the pre-test, a two-task computer-based activity was designed by the research group: this activity shared the same characteristics as the pre-test activity but it was focused on the Moon. Like in the pre-test activity, the content of this activity was new to the students and the activity was also solved individually, rather than in group The post-test activity also contained the same two types of tasks: task 1: *Fact-finding* task; task 2: *No-fact-finding* task.

3.2.3. Data analysis

From a quantitative point of view, students' success in finding right information on the web as well as answering questions correctly was evaluated with the answers' degree of correctness. The students' answers were scored in a binary fashion as either correct (1) or incorrect (0). The independent variables *Fact-Finding*, *No-Fact-Finding* and *Total scores* were computed and evaluated.

3.2.4. Results

Pre-test and Post-test scores

Table 3 shows the mean scores and standard deviations for the experimental and control groups for the three variables analyzed: (a) *Fact-Finding*, (b) *No-Fact-Finding*, and (3) Total.

Pre-Test Scores

The total Pre-test performances were low for both groups. An independent t-test was conducted to determine whether or not there were differences between experimental and control groups at the beginning of the intervention. Results revealed no significant differences at the 95% significant level for the three variables [Pre-test *Fact-Finding*: $t(125) = 1.103$, $p = 0.272$; Pre-test *No-Fact-Finding*: $t(125) = 1.233$, $p = 0.220$; Pre-test *Total*: $t(125) = 1.325$, $p = 0.187$]. Thus, control and experimental samples were equivalent at the beginning of the experiment.

Post-Test Scores

An independent t-test was also conducted to determine differences between control and experimental groups after the intervention. Results showed no differences for the *Fact-Finding* variable [Post-test *Fact-Finding*: $t(125) = -1.116$, $p = 0.266$]. However, significant differences were found for the *No-Fact-Finding* and *Total* variables [Post-test *No-Fact-Finding*: $t(125) = -3.144$, $p = 0.002$; Post-test *Total*: $t(125) = -2.383$, $p =0.019$]. Experimental students outperformed their counterparts in the control group, for the three variables analyzed.

Table 3. Pre-test and post-test mean scores and standard deviations for control and experimental groups.

	Possible Points	CONTROL GROUP				EXPERIMENTAL GROUP			
		Pre test		*Post test*		*Pre test*		*Post test*	
		M	SD	M	SD	M	SD	M	SD
Fact-Finding	**14**	**10.97**	**2.38**	**9.15**	**2.10**	**10.45**	**2.48**	**9.65**	**2.38**
No-Fact-Finding	**7**	**2.04**	**1.55**	**2.37**	**1.34**	**1.69**	**1.41**	**3.36**	**1.75**
Total	**21**	**13.01**	**3.34**	**11.53**	**2.77**	**12.15**	**3.41**	**12.99**	**3.36**

n(Control)= 39 ; n(Experimental)=88

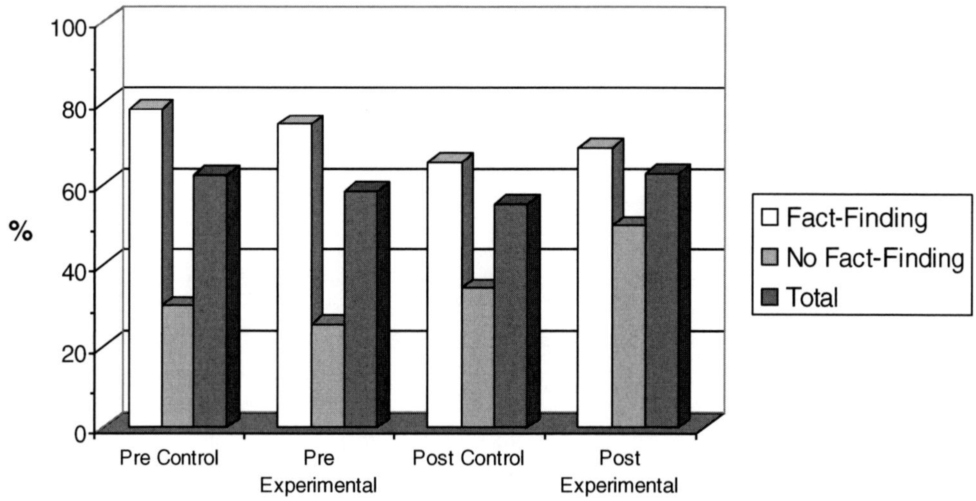

Figure 8. Percentage of correct answers by type of task: pre-test and post-test comparisons for Experimental and Control groups.

Pre-test and post-test comparisons

Paired-sample t-tests were conducted to evaluate differences before and after the intervention for the control group and the experimental group, respectively. In the control group, significant differences were assessed for the *Fact-Finding* and the *Total* variables [*Fact-Finding*: t(38)=5.257; p=0.000; *Total*: t(38)=3.108; p=0.004].

In the experimental group, significant differences were found for the three variables [*Fact-Finding*: t(87)=2.873; p=0.005; *No-Fact-Finding*: t(87)=-8.904; p=0.000; *Total*: t(87)= -2.467; p=0.016]. Total scores in the post-test were higher than in the pre-test; indicating a positive effect of the experiment where the instructional approach was implemented

Histograms of Figure 8 represent the percentage of the students' population who answered correctly each type of task for the Pre-test and the Post-test in the Experimental and Control groups.

Figure 8 revealed that the students' degree of correctness depended on the type of information students searched on the web. In our case, the students' degree of correctness was higher when they were asked to locate specific data (*Fact-Finding*) than when they were asked to search for conceptual information and arguments to defend a position about a topic (*No-Fact-Finding*).

From Figure 8, the percentage of correct answers in the Pre-test *Fact-Finding* questions was 78.39% for the control group (M=10.97, SD=2.38) and 74.68% for the experimental group (M=10.45, SD=2.48). When examined by the Pre-test *No-Fact-Finding* questions, the percentage of correct answers decreased to 29.14 % for the control group (M=2.04, SD=1.55) and 24.14% for the experimental group (M=1.69, SD=1.41). No significant differences were detected between control and experimental samples (p>0.05).

The most relevant point of Figure 1 for the post-test scores is that there is a significant increase in the percentage of correct answers for the Post *No-Fact-Finding* from the control group, 34.43% (M=2.37, SD=1.34) to the experimental group, 49.84% (M=3.36, SD=1.75);

indicating that students in the experimental group outperformed significantly ($p=0.002$) those in the control group for the *No-Fact-Finding* questions. No differences were observed for the *Fact-Finding* variable ($p>0.05$).

3.3. Discussion

Pre-test

The first research question focused on determining if secondary students have difficulties in finding and integrating web information to construct knowledge in complex science topics at the beginning of the intervention. While no significant difference was detected between control and experimental groups, results revealed that the pre-test performance was low for both groups. A possible reason for the overall weak pre-test performance may stem from the difficulty of the content. Another reason may be the complex nature of the skills involved in finding and integrating web information to master the content. It is important to point out that the students' performance with *Fact-Finding* questions was better than that with *No-Fact-Finding* questions. These results revealed that secondary education students have difficulties in finding information on the web, understanding its content, and selecting the right information to answer *No-Fact-Finding* questions. In turn, it indicated that students get lost when searching for web information to answer complex science open tasks (*No-Fact-Finding*). These tasks require students to search for information, find the right one, read it carefully, select the information effectively and critically, and synthesise it.

These results are consistent with previous findings from other researchers who also noted such difficulties among students (Navarro-Prieto, Scaife & Rogers, 1999; Wallace, Kupperman, Krackcik & Soloway, 2000; Brush & Saye, 2001; Chui, 2002; Azevedo, Cromley & Seibert, 2004b; Ruthven, Henessy & Deaney, 2005).

Furthermore, these results justify the interest of implementing our instructional design to help students' master conceptual understanding of complex science topics and answer meaningful *No-Fact-Finding* questions. Web users need specific skills ranging from orientation and search planning to selecting and evaluating information in order to benefit from the many information resources provided by the WWW (Brand-Gruwel, Wopereis, & Vermetten, 2005). It seems rather unlikely that students may acquire these skills and knowledge prerequisites on the fly by engaging in web searches on their own.

Post-test

The second research question inquired whether or not the same difficulties observed in the pre-test persisted once the intervention was concluded. Results indicated that students who followed the instructional approach performed significantly better than students who did not. It may be concluded that the instructional approach assisted students in finding the right information, integrating and elaborating it, and eventually giving a personal opinion (*No-Fact-Finding* tasks).

The results of our study clearly demonstrated the effectiveness of the instruccional web-based learning designed. The support that students received in asking questions, in developing efficient searching and managerial strategies and in the construction of new understanding was efficient in improving the students' web searching skills and in constructing new knowledge in Science.

These findings are similar to those obtained in other studies. In this line, Hoffman, Wu, Krajcik & Soloway (2003) describe how students move to high levels of content understanding and learning when they interact and learn from an on-line learning environment designed to support question asking, resource gathering and the construction of new understandings. Gerjets & Hellenthal-Schorr (2008) report on the efficiency of user-oriented training for information search on the web. In this study the students increased their specific web searching and managerial skills. The training also had the advantages of higher durability of acquired competencies and the potential to generalize the contents acquired.

Task characteristics

Finally, the third research question focused on studying the effect of the instructional task characteristics on the students' success in finding and using web information for learning purposes on science.

The results of this study suggest that the type of task and cognitive actions required to answer properly the curricular question are critical to explain the students' degree of correctness. Performance decreased when students were asked to develop complex managerial strategies of the information found in the web to answer *No-Fact-Finding* tasks. This may be explained by the fact that students had to develop complex cognitive processes to assess, comprehend and integrate digital information.

From our data in the pre-test, only around 27 % of the students answered correctly the *No-Fact-Finding* questions while around 70 % did it for the *Fact-Finding*. Previous works drew similar conclusions. Mayordomo & Garcia-Mila (2005) reported scores of 45% of incorrect answers in finding concrete data in hypertext conditions. In another context, Dreher (1993) found that more than 55% of students had not located the correct answer to questions in a textbook chapter in spite of the fact that they had inspected the page of the text containing the answer. The explanation of the students' failure is that efficient search of text or hypertext involves good planning (i.e. selecting the "best" sections of text) and efficient and monitoring extraction (i.e. trying not to miss information) (Dreher, 1993; de Jong & Van der Hulst, 2002; Rouet *et al.* 2001).

When analysing the experimental group, significant differences were detected between the pre-test and the post-test. An important increase in the percentage of correct answers was detected for the experimental group between the pre-test (25.49%) and the post-test (49.84%).

Results confirmed that designing instructional approaches which combine the use of computers and Internet with IBL and procedural guidance can improve learning of complex science topics. As indicated in previous sections *No-Fact-Finding* questions are more complex than *Fact-Finding* ones and require students to master concept understanding and develop coherent and generative scientific knowledge.

The significant increase in solving *No-Fact-Finding* questions shows that experimental students increased the skills to solve more efficiently information problem solving; all these skills were taught during the instructional process (Hill & Hannafin 1999; Lazonder, 2000; Brand-Gruwel, Woperis & Vermetten (2005).

4. CONCLUSIONS

Both our study and the studies reported in this paper highlighted some difficulties of secondary students to use World Wide Web as a learning tool in Science. These studies demonstrate the necessity to support students in acquiring skills for a more competent search and management of information on the Internet. Accordingly, it can be claimed that a specialized instruction for information search on the web is an essential educational task in order to enable students to autonomously use a wide range or information resources and to face the requirements of modern information society.

In this context, the necessity to generate instructional environments which allow to integrate the World Wide Web as a learning tool in the curriculum and enhancing metacognitive domain and system knowledge appeared critical to improving students' web performance (Land & Greene, 2000). To reach this ambitious objective, from our point of view, the design of these instructional environments in Science Education need to take into account the following four interrelated instructional components:

a) Promoting inquiry web learning. An important aspect of inquiry and a crucial issue in developing conceptual understanding is generating explanations and theories for the phenomena being investigated (Perkins, Crismond, Simmons & Unger, 1995; Scardamalia & Bereiter, 1994). In the computer-based activities of our project, information-seeking and web-based information understanding were presented as problems to solve for students immersed in a learning environment designed to support asking questions, gathering resources and constructing new understanding. To do so, we have used WebQuest methodology, as reported by Dodge (1995). WebQuest methology encourages students to think and plan the different phases to solve a task. In our project we have stressed the importance of the process section. In this section we scaffolded the students' resolution of a problem, in particular, the development of efficient searching and managerial strategies which could promote content understanding.
b) Using scaffolding to support the learning process (Palinscar & Brown, 1984; Paris, Wixson, & Palinscar, 1986; Rosenshine & Meister, 1992). This study viewed scaffolding as a process of providing decreasing amounts of support to help students bridge the gap between their current abilities and the intended goal of instruction that allows students to participate at ever-increasing levels of competence (Rosenshine & Meister, 1992). Scaffolding materials appear in our work in the form of procedural guidelines that support the students' resolution process. These scaffolded instructions allowed students to gain appropriate experience and skills to increase their cognitive capabilities toward the task.
c) Teaching specific skills, procedures, and strategies to search and manage web information for learning purposes, and these strategies were implemented in all the activities. At present, in the computer-based activities we are working on, special attention is paid to the following five main strategies: strategies to search, elaborate and organize web information; strategies to produce and construct one's own and personalised information and strategies to monitor and regulate all the problem-solving information process.

d) Promoting the students' access and work with different educational web resources: e.g. simulators, applets, hypertext, encyclopaedia. In this line of work, our research developed, implemented and evaluated an instructional environment for integrating the World Wide Web as a learning tool in the secondary education science curriculum. From our point of view, the better results obtained in the post-test by the experimental group students show that the intervention approach enhances the students' development of more efficient searching and managerial web-information strategies and show a positive path in the design of instructional processes capable to enhance the students' digital literacy.

Taking into account the conclusions presented above, our research group is currently continuing on the work to develop, implement and evaluate a scaffolding instructional process which integrates the use of the Internet to learn different curricular contents. This project will focus on improving and evaluating the evolution of searching and managerial strategies as a tool for students to take full advantage of web information for learning purposes.

ACKNOWLEDGMENTS

This research was funded by the Ministerio de Ciencia y tecnología of the Spanish Government (project numbers: BSO2003-05409, SEJ2006-12110 and EDU2009-11656). The author would like to thank the high-school teachers and their students for their participation in this study.

5. REFERENCES

Azevedo, R., Cromley, J. G. & Seibert, D. (2004a). Does adaptive scaffolding facilitate students' ability to regulate their learning with hypermedia? *Contemporary Educational Psychology*, *29*, 344-370.

Azevedo, R., Cromley, J. G. & Seibert, D. (2004b). The role of self-regulated learning in fostering students' conceptual understanding of complex systems with hypermedia. *Journal of Educational Computing Research*, *30(1)*, 87-111.

Azevedo, R., Cromley, J. G., Winters, F. I., Moos, D. C. & Greene, J. A. (2005). Adaptive Human scaffolding facilitates adolescents' self-regulated learning with hypermedia. *Instructional Science*, *33*, 381-412.

Brand-Gruwel, S. Wopereis, I. & Vermetten, Y. (2005). "Information problem solving by experts and novices: analysis of a complex cognitive skill". *Computers in human behaviour*, *21*, 487-508.

Brush, T. & Saye, J. (2001). The Use of Embedded Scaffolds with Hypermedia-Supported Student-Centered Learning. *Journal of Educational Multimedia and Hypermedia*, *10*, 4, 333-356.

Chang, C. Y. (2001) Comparing the Impacts of a Problem-Based Computer-Assisted Instruction and the Direct-Interactive Teaching Method on Student Science Achievement. *Journal of Science Education and Technology, 10 (2)* 147-153.

Chi, M. T. H:, Siler, S., Jeong, H., Yamauchi, T. & Hausmann, R. (2001) Learning from Human Tutoring, *Cognitive Science, 25*, 471-534.

Chui, M. (2002) I still haven't found what I'm looking for: Web searching as query refinement. Ph dissertation. Indiana University. US.

Coye, R. W. & Stonebraker, P. W. (1994) The Effectiveness of Personal Computers in Operations Management Education. *International Journal of Operations & Production Management, 14*, 35-46.

De Jong, T. & Van der Hulst, A. (2002). The effects of graphical overviews on exploratory behaviour and knowledge acquisition in hypertext environments. *Journal of Computer Assisted Learning, 18*, 219-232.

Dodge, B. (1995). "WebQuests: A Technique for Internet-Based Learning." *Distance Educator 1(2)*, 10-13.

Dodge, B. (1997) *Some Thoughts about WebQuest* (accessed 3/14/07). http://webquest.sdsu.edu/about_webquests.html

Dreher, M. J. (1993). Reading to locate information: societal and educational perspectives. *Contemporary Educational Psychology, 18*, 129-138.

Eisenberg, M. B, & Berkowitz, R. E. (1990). *Information-problem solving: the big six skills approach to library and information skills instruction.* Norwood, NJ: Ablex.

Ferguson, N. H, & Chapmen, S. R. (1993) Computer-Assisted Instruction for Introductory Genetics. *Journal of Natural Resources and Life Science Education, 22*, 145-152.

Greene, B. & Land, S. (2000) A qualitative analysis of scaffolding use in a resource-based learning environment involving the world wide web. *Journal of Educational Computer Research, 23 (2)*, 151-179.

Gerjets, P. & Hellenthal-Schorr, T. (2008). Competent information search in the World Wide Web: Development and evaluation of a web training for pupils. *Computers in Human Behavior, 24*, 693-715.

Guinee, K., Eagleton, M. & Hall, T. (2003). Adolescents' Internet search strategies: drawing upon familiar cognitive paradigms when accessing electronic information sources. *Journal of Educational Computing Research., 29 (3)*, 363-374.

Hill, J. & Hannafin, M. (1999) Cognitive strategies and learning from the World Wide Web. *Educational Technology Research and Development, 45(4)*, 37-64.

Hoffman, J., Wu, H., Krajcik, J. & Soloway, E. (2003). The nature of middle school learners' science content understandings with the use of on-line resources. *Journal of Research in Science Teaching, 40 (3)*, 323-346.

Hsieh, C. T. & Lin, B. (1998) Educational Value of the Internet: Students' Perceptions, Communications of the ICISA. Fall, 25-31.

Kao, M. & Lehman, J. (1997) Scaffolding in a computer-constructivist environment for teaching statistics to college learners. Paper presented at the annual meeting of the American Educational Research Association, Chicago, IL.

Land, S. & Greene, B. (2000) Project-based learning with the Worl Wide Web: A qualitative study of resource integration. *Educational Technology Research and development, 48(3)*, 61-78.

Lazar, J. (2003). Training novice users in developing strategies for responding to errors when browsing the web. *International Journal of Human-Computer Interaction, 15 (3)*, 361-377.

Lazonder, A. W (2000). Exploring novice users' training needs in searching information on the WWW. *Journal of computer Assisted Learning, 16 (4)*, 326-335.

Lee, S. C. (2001) Development of Instructional Strategy of Computer Application Software for Group Instruction. *Computers & Education, 37*, 1-9.

Lim, B. (2004) Challenges and Issues in Designing Inquiry on the Web. *Bristish Journal of Educational Technology, 35 (5)*, 627-643.

Martinello, M. & Cook, G. (2000) *Interdisciplinary Inquiry in Teaching and Learning* (2nd Edition), Pearson Education, NJ: Upper Saddle River

Mayordomo, R. & Garcia-Mila, M. (2005). The effect of type of access to digital information on data searching and text comprensión. Proceedings of 11th Bienial EARLI conference. Integrating Multiple Perspectives on Effective Learning Environments. Cyprus, August 23-27.

McReary, C. L., Golde, M. F. & Koeske, R. (2006) Peer Instruction in the General Chemistry Laboratory: Assessment of Student Learning. *Journal of Chemical Education, 83 (5)*, 804-810.

Navarro-Prieto, R., Scaife, M. & Rogers, Y. (1999). *Cognitive strategies in web searching. Proceedings of 5th conference on Human Factors & Web.* June. Maryland.

Palinscar, A. M. & Brown, A. L. (1984). Reciprocal teaching of comprehension fostering and comprehension-monitoring activities. *Cognition and Instruction, 2*, 117-175.

Paris, S. G., Wixson, K. K. & Palinscar, A. M. (1986). Instructional approaches to reading comprehension. In E. Z. Rothkof (Ed.), *Review of Research in Education.* Washington D.C.: American Educational Research Association.

Pedersen, S. & Liu, M. (2002a) The transfer of problem-solving skills from a problem-based learning environment: The effects of modeling an expert's cognitive processes. *Journal of Research on Technology in Education, 35(2)*, 303-320.

Pedersen, S. & Liu, M. (2002b) The effects of modeling expert cognitive strategies during problem-based learning. *Journal of Educational Computing Research, 26(4)*, 353-380.

Pelgrum, W. J. (2001). Obstacles to the integration of ICT in education: results from a worldwide educational assessment, *Computers & Education, 37*, 163-178.

Perkins, D. N., Crismond, D., Simmons, R., & Unger, C., (1995). "Inside understanding". In D. N. Perkins, J. L. Schwartz, M. M. West, & M. S. Wiske (eds.), *Software goes to school: Teaching for understanding with new technologies.* New York, NY: Oxford University Press, 70-87

Pifarré, M., Gòdia, S. & Martorell, I. (2006). Analyses of cognitive strategies used in web searching by secondary students: Implications for instructional design. In G. Clarebout & J. Elen (Eds.). *Avoiding simplicity, confronting complexity. Advances in studying and designing (computer-based) powerful learning environments.* Rotterdam: Sense Publishers.

Pifarré, M., Sanuy, J., Vendrell, C. & Godia, S. (2008). *Internet en la Educación Secundaria. Pensar, buscar y construir conocimiento en la red.* Lleida: Milenio.

Pifarré, M., Sanuy, J., Vendrell, C., Godia, S. & Martorell, I. (2007). *Project. Internet in Secondary Education*: http://webquest.udl.es

Powell, J. V., Aeby, V. G. & Jr., Carpenter-Aebyc, T. (2003) A Comparison of Student Outcomes with and without Teacher Facilitated Computer-Based Instruction. *Computers & Education*, *40*, 183-191.

Prince, M. J. & Felder, R. M. (2006). Inductive Learning and Learning Methods: Definitions. Comparisons, and Research Bases. *Journal of Engineering Education*, *April* 1-16.

Rosenshine, B. & Meister, C. (1992). The use of scaffolds for teaching higher-level cognitive strategies. *Educational Leadership*, *49(7)*, 26-33.

Rouet, J. F., Vidal-Abarca, E., Bert-Erboul, A. & Millogo, V. (2001). Effects of information search tasks on the comprensión of instructional text. *Discourse Processes*, *3*, 163-186.

Rowe, G. W. & Gregor, P. (1999) A Computer Based Learning System for Teaching Computing: Implementation and Evaluation, *Computer & Education*, *33*, 65-76.

Rubin, S. (1996) Evaluation ad Meta-Analysis of Selected Research Related to the Laboratory Component of Beginning College Level Science, PhD. Dissertation, Temple University.

Ruthven, K., Henessy, S. & Deaney, R. (2005) Incorporating Internet resources into classroom practice: pedagogical perspectives and strategies of secondary-school subject teachers, *Computers & Education*, *44*, 1-34.

Ryder, R. J. & Graves, M. F. (1997). Using the internet to enhance student's reading, writing, and information-gathering skills. *Journal of Adolescent & Adult Literacy*, *40(4)*, 244-254.

Scardamalia, M. & Bereiter, C. (1994). Computer support for knowledge-building communities. *The journal of the learning sciences*, *3*, 265-283.

Shute, Y. & Bonar, J (1996) Intelligent Tutoring Systems for Scientific Inquiry Skills, ERIC Document Reproduction Service No. *299*, 134.

Shymansky, J., Hedges, L. & Woodworht, G. (1990). A Reassessment of the Effects on Inquiry-Based Science Curricula of the 60's on Student Performance. *Journal of Research in Science Teaching*, *27 (2)*, 127-144.

Simons K. D. & Klein J. D. (2007). The Impact of Scaffolding and Student Achievement Levels in a Problem-based Learning Environment. *Instructional Science*, *35*, 41-72.

Slotta, J. (2002). Designing the Web-based Inquiry Science Environment (WISE). *Educational Technology.*, *42*, 15, Cambridge.

Smith, D. (1996). *A Meta-Analysis of Students Outcomes Attributable to the Teaching of Science as Inquiry as Compared to Traditional Methodology*, PhD. Dissertation, Temple University, Department of Education.

Tabatabai, D. (2004). Modeling information-seeking expertise on the web. PhDissertation. Mcgill University. Canada.

Tjaden, B. J. & Martin, C. D. (1995) Learning Effects of Computer-Assisted Instruction on Collage Students. *Computer & Education*, *24*, (4) 221-277.

Tsai, C. C. & Chou, C. (2002) Diagnosing Students' Alternative Conceptions in Science, *Journal of Computer Assited Learning*, *18*, 157-165

VanLehn, K., Siler, S., Murray, C., Yamauchi, T. & Bagget, W. (2003) Why do only some events cause learning during human tutoring? *Cognition & Instruction*, *21(3)*, 209-250.

Wallace, R., Kupperman, J., Krackcik, J. & Soloway, E. (2000) Science on the Web: Students Online in a Sixth-grade Classroom. *The Journal of the Learning Sciences*, *9 (1)*, 75-104.

In: Computer-Assisted Teaching: New Developments ISBN: 978-1-60876-855-4
Editors: B. A. Morris, G. M. Ferguson, pp. 93-115

Chapter 4

USING COMPUTER-ASSISTED TEACHING TO PROMOTE CONSTRUCTIVIST PRACTICES IN TEACHER EDUCATION

Rebekah K. Nix[1*] and Barry J. Fraser[2**]
[1]The University of Texas at Dallas, USA.
[2]Curtin University of Technology, Perth, Australia.

ABSTRACT

In this chapter, we report how the incorporation of information and communications technology (ICT) into a teacher professional development program fostered constructivist teaching/learning practices in school classrooms. Changing the teachers' learning environment at the university level, based partly on computer-assisted teaching, fostered similar changes in their students' middle-school learning environments. The Constructivist Learning Environment Survey (CLES), which provided a critical scaffold for design and delivery, was administered to 845 students of 17 teachers. When compared to other classrooms within the same middle schools, these science classrooms were perceived to be significantly more positive in terms of the CLES scales of Personal Relevance, Shared Control, Critical Voice, and Student Negotiation. Qualitative data from observations and in-depth case studies provided further insight into the ways in which the ICT-enhanced program changed the classroom practice of experienced science teachers. Four assertions regarding the implementation of constructivism were identified relating to teacher collaboration, efficacy, understanding, and facility.

INTRODUCTION

New approaches to teaching and learning are being developed in response to the information revolution. Combined with the political and social attention to education on a

* Corresponding authors: E-mail: rnix@utdallas.edu
** E-mail: B.Fraser@curtin.edu.au

global scale, the infusion of information technology into the classroom has supported similarly rapid and important advances in learning environments research (Fraser, 2001, 2007; Khine & Fisher, 2003) and constructivist pedagogy (von Glaserfeld, 1995). Unfortunately, few studies offer concrete solutions for such issues as teacher preparation, transfer of meaningful learning, and effective use of teaching tools for achieving sustained positive change within today's educational system. "If educational innovations are to succeed, they must take a more realistic view of the realities of classroom life than have some past curricular projects" (Geelan, 1997, p. 4).

Our study focused on an innovative professional development program that was designed to provide a tangible opportunity for teachers to gain organized knowledge to guide practical changes in their school classrooms (Nix & Ledbetter, 2005). Based on the Integrated Science Learning Environment (ISLE) model, a variety of scientific and educational technologies were employed to merge theory and practice with science and education through effective applications of information technology to create a rich learning environment (Nix, 2002). An integrated approach to science education emphasized interconnections among teachers' current interests and areas of specialization. Coursework was structured to help teachers to design individualized plans for implementing new experiences into their classroom teaching.

Because today's teachers were students in a totally different world, there are important implications for realizing the potential of constructivist practice. The Committee on Undergraduate Science Education (1997) notes that research indicates that teachers teach in a manner consistent with their own way of learning (Shulman, 1990; Tobin, Tippins, & Gallard, 1994). Evolving national, state, and district reforms aimed at developing consistency and continuity across the curriculum are likely to help to advance the application and integration of new teaching tools and resources. But, bringing teachers, technology, students, and learning together requires a new model of education that is suitable for tomorrow's classrooms. As Rillero (1993) noted with regard to teacher education programs, if thorough training has direct relevance, it will be "a natural development for the materials to pass from the teacher's hand into the student's hands" (p. 15). The teacher education program investigated in our study attempted to improve classroom learning environments by basing teacher education on constructivist principles.

Building and expanding upon previous preliminary research (Nix, Fraser, & Ledbetter, 2005a, 2005b), this study involved evaluating the ISLE model using teacher and student perceptions of dimensions of the learning environment as dependent variables. The overall research design involved assessing the program in terms of promoting more constructivist classroom learning environments, improving teachers' pedagogical expertise, and advancing teachers' content knowledge. The study reported in this article focused on the educational impact of the emergent model.

This article reports research that shared some common elements with a previous evaluation of the ISLE model (Nix et al., 2005a), but also extended previous work in important ways. Whereas the previous study focused on the implementation of the ISLE model within a summer short course for teachers, the new study focused on an implementation of the same ISLE model that spanned three semesters. Entirely different samples of teachers and school students were involved in the two different studies, although they both were drawn from the same population of north Texas teachers and the same graduate program. Whereas our previous study employed the original longer version of the Constructivist Learning Environment Survey (CLES) to assess classroom environment, the present study made use of a newer and shortened version of the CLES developed by Johnson and McClure (2004).

BACKGROUND

Integrated Science Learning Environment (ISLE)

Experienced science teachers demonstrate wide individual differences in content knowledge and pedagogical skills that impact on the classroom learning environment that they foster for their students. Therefore, to realize a shift in teaching practice, "as science educators, we need to be able to support our personal efforts and teachers' efforts...(by)...trying to model such practices in our own classrooms" (Moscovici, 1999, ¶ 19).

The ISLE model holistically combines a variety of approaches that can lead to the development of a more constructivist classroom learning environment. Extending continuously across the intellectual, physical, and emotional boundaries of the university, field trip, and class/school learning environments, it supports and encourages classroom teachers in the implementation of new technologies and teaching strategies (Nix, Ledbetter, & Fraser, 2001). The purpose of the model is to bring about personal growth through activities that place science-related content into perspective and to apply the principles of collaborative problem solving in a real-world setting. With successful transfer (Foxon, 1993), the same techniques used to deal with issues in the university setting are applied to integrating new knowledge and expertise in school classrooms.

Three key aspects distinguish the ISLE model from traditional teacher development programs. First, ISLE models the integration of information technologies into the university classroom and curriculum, as they might be implemented in the school classroom. By actually experiencing the appropriate and effective use of information technologies in educational practice, teachers can appreciate the value of new tools and resources. Second, ISLE encompasses the field trip, as well as the university and public/private school classroom milieus. By focusing on the common element, the individual, experience can be internalized and thereby naturally transferred among the physical settings. Third, ISLE seamlessly presents information technology as a means to an end, not the end itself. By selecting and applying appropriate tools and resources, the benefits (rather than the implementation challenges) can be maximized.

Field of Learning Environments

Our study drew on and contributed to the field of classroom learning environments (Fisher & Khine, 2006; Fraser, 1994, 1998b, 2007; Fraser & Walberg, 1991; Goh & Khine, 2002; Khine & Fisher, 2003) and is consistent with a long-standing tradition in this field of obtaining quantitative information through the administration of established and economical questionnaires which assess students' perceptions of their classroom learning environment (Fraser, 1998a). In past research, student perceptions of classroom environment have frequently been used as predictor variables to establish consistent relationships between the nature of the classroom environment and student cognitive and affective outcomes (Goh, Young, & Fraser, 1995; Haertel, Walberg, & Haertel, 1981; McRobbie & Fraser, 1993). Furthermore, research involving a person-environment fit perspective has shown that students achieve better where there is greater congruence between the actual classroom environment and that preferred by students (Fraser & Fisher, 1983).

Another common application of classroom environment assessments has been their use as dependent variables in evaluating educational innovations (Lightburn & Fraser, 2007; Maor & Fraser, 1996; Martin-Dunlop & Fraser, 2008; Ogbuehi & Fraser, 2007; Spinner & Fraser, 2005). For example, in comparison with a control group, students using Australian Science Education Project materials perceived their classrooms as being more satisfying and individualized and having a better material environment (Fraser, 1979). Classroom environment variables differentiated revealingly between curricula, even when various outcome measures showed negligible differences. When Maor and Fraser (1996) incorporated a classroom environment instrument within an evaluation of the use of a computerized database, it was found that students perceived that their classes became more inquiry-oriented during the use of the innovation. Studies in Singapore used classroom environment measures as dependent variables in evaluations of computer-assisted learning (Teh & Fraser, 1994) and computer application courses for adults (Khoo & Fraser, 2008). Zandvliet and Fraser (2004) focused on the effects of internet technology in Canadian and Australian high schools, whereas Dryden and Fraser (1998) used learning environment dimensions in assessing the impact of a statewide initiative aimed at changing science instruction in Texas.

The study reported in this chapter is noteworthy because it used classroom environment variables as process criteria of effectiveness in evaluating an innovative teacher education program that incorporated ICT. The instrument used, the Constructivist Learning Environment Survey (CLES), is discussed in the next section.

Constructivist Learning Environment Survey (CLES)

The principles of constructivist teaching guided the design of ISLE-based programs, providing a common thread throughout the coursework and the formative and summative evaluation of the program. In response to the need to assess innovative classroom environments, the CLES was developed with a psychological view of learning that focused on students as co-constructors of their own knowledge (Taylor & Fraser, 1991; Taylor, Fraser, & Fisher, 1997). A unique aspect of the CLES is that items from the same scale are grouped together. The original 30-item version of the CLES contains six items with a five-point frequency response scale (5 = Almost Always, 4 = Often, 3 = Sometimes, 2 = Seldom, and 1 = Almost Never) in five scales:

Personal Relevance (relevance of learning to students' lives)
Uncertainty of Science (provisional status of scientific knowledge)
Critical Voice (legitimacy of expressing a critical opinion)
Shared Control (participation in planning, conducting and assessing of learning)
Student Negotiation (involvement with other students in assessing viability of new ideas).

Numerous past studies attest to both the validity of the CLES in numerous countries and to its usefulness in various research applications. In a cross-cultured study involving use of the CLES in two languages (English and Mandarin) at the junior high school level, 1879 Taiwanese science students in 50 classes were compared with 1081 Australian science students in 50 classes. The study provided strong support for the validity of the CLES (factorial validity, reliability, and ability to differentiate between classrooms), established associations between student attitudes and classroom environment, and revealed that Australian science classes were perceived as more constructivist than Taiwanese science classes in terms of several CLES scales (especially Critical Voice and Student Negotiation).

When Kim, Fisher, and Fraser (1999) translated the CLES into the Korean language and crossvalidated it with a sample of 1083 students in 24 Grade 10 science students, results supported the factor structure and reliability of the Korean version, revealed statistically significant relationships between classroom environment and students' attitudes to science, and confirmed that students exposed to a new curriculum perceived a more constructivist learning environment than did students who had not been exposed to this curriculum.

In two other studies, Korean researchers collaborated with an American colleague in research involving the use of a Korean version of the CLES. As part of an action research project involving creating constructivist learning environments in Grade 11 earth science classes, 136 Korean students responded to the CLES several times in a longitudinal study of the development of constructivist classrooms and students' attitudes (Oh & Yager, 2004). Not only were there improvements in CLES scores over time, but students' attitudes to science became more positive as their classrooms became more constructivist. Cho, Yager, Park, and Seo (1997) used this version of the CLES with 70 Korean high school teachers who visited the University of Iowa for professional development programs. When the CLES was administered three times (at the beginning and the end of workshops and three months later) to evaluate the program in terms of the development of teachers' constructivist philosophies, initial improvements in CLES scores were found, but these gains tended not to be retained over a longer time period.

In South Africa, Aldridge, Fraser, and Sebela (2004) used the CLES with 1864 students in 43 Grade 4–9 mathematics classes. They crossvalidated the CLES for use in South Africa (in terms of factor structure, reliability, and ability to differentiate between classrooms), and reported the success of a subsample of teachers in using feedback from the CLES in guiding action-research attempts at improving the constructivist orientation of their classrooms.

In Australia, Taylor et al. (1997) included the CLES among a random sample of 494 13-year-olds in 41 Grade 8 and 9 classes in 13 schools as an optional component of the Third International Mathematics and Science Study (TIMSS). For this Australian sample, the CLES displayed sound factorial validity and internal consistency reliability.

In Florida, two studies employed modified versions of the CLES in evaluating educational programs. When Peiro and Fraser (in press) used simplified English and Spanish versions of the CLES with 739 Grades K–3 science students, the CLES was crossvalidated for use in either English or Spanish with this sample of young children. Also the effectiveness of interventions aimed at creating a more constructivist learning environment was supported by statistically significant and large changes (around two standard deviations) between pretest and posttest on the two CLES scales of Personal Relevance and Student Negotiation. In the other study in Florida, Spinner and Fraser (2005) used the CLES with two separate samples of 53 and 66 fifth-grade students undertaking an innovative mathematics program called the

Class Banking System (CBS). As well as crossvalidating the CLES, these researchers reported that, relative to non-CBS students, CBS students experience more favourable pre-post changes on most of the dimensions of the CLES.

In Texas, the CLES was used in monitoring the success of a systemic educational reform effort that included more emphasis on constructivist teaching and learning (Dryden & Fraser, 1998). Although the study reported strong evidence to support the validity and reliability of the CLES with a sample of 1600 Grade 9–12 science students, a longitudinal study over several years painted a disappointing picture of little success in moving towards more constructivist science instruction in Texas.

Cannon (1995) used the CLES in evaluating an elementary science methods course which was based upon constructivist epistemology and fostered various constructivist teaching and learning strategies. When the CLES was administered to 43 preservice elementary teachers (mainly females) at an American university at the end of the course, CLES scales exhibited satisfactory reliability. However, median scores on CLES scales were lower than anticipated by the researcher-instructor. Nevertheless, feedback from the CLES was productive in identifying areas within the methods course that were less consistent with constructivist epistemology and in motivating the researcher-instructor to re-examine and modify classroom practices.

Beck, Czerniak, and Lumpe (2000) used the five constructs of the CLES (Personal Relevance, Scientific Uncertainty, Critical Voice, Shared Control, and Student Negotiation) in a study of teachers' beliefs regarding implementing constructivism in their classrooms. Two samples of 47 and 203 teachers in Ohio were used to elicit salient beliefs about implementing the components of constructivism using Ajzen's (1985) theory of planned behaviour. A modified version of the CLES provided a measure of teachers' self-reported implementation of issues related to constructivism. Beck and colleagues reported data to support the reliability of the CLES and concluded that, if teaching in a constructivist fashion is desired in schools, then teachers' beliefs about this behaviour must first be considered.

Harwell, Gunter, Montgomery, Shelton, and West (2001) reported the use of the CLES in the US in a collaborative action research endeavour between a regional university and a local school (Grade 6 level) to monitor alignment of classroom learning activities with a constructivist viewpoint while integrating technology into the curriculum. Teacher logs, teacher interviews and field notes from team discussion groups and classroom observations provided further understanding of interactions in the classroom. Harwell and colleagues reported satisfactory alpha reliability coefficients for all CLES scales for a small sample of approximately 60 students, but found no significant changes in student perceptions of the classroom learning environment over the duration of the academic year. Interpretation of results led teachers to construct a new set of questions and a new plan of action to bring their classroom learning environments into closer alignment with a constructivist perspective for teaching and learning.

Johnson and McClure (2004) administered the CLES to 290 science teachers to provide insights into classrooms in Minnesota. Teacher interviews, analysis of videotaped lessons, and teacher and student versions of the CLES were used to gather data that led to assertions about teacher knowledge and beliefs, teaching performance, and a comparison of knowledge and beliefs to teaching performance.

In our previous study involving the use of the original 30-item CLES among 1079 students in 59 classes in North Texas, we reported strong factorial validity and reliability (Nix et al., 2005a). When Johnson and McClure (2004) used the same original 30-item version in the USA with 290 upper-elementary, middle-school and high-school teachers and preservice teachers, they also reported strong factorial validity and reliability. Nevertheless, Johnson and McClure developed a shorter and modified 20-item version of the CLES containing the same five scales. For a different sample of teachers and students at the upper-elementary, middle-school and high-school levels, Johnson and McClure reported that the new and more economical version of the CLES exhibited strong internal consistency reliability. Consequently, we decided to use Johnson and McClure's short version of the CLES in the study reported in this article.

RESEARCH METHODS

Evidence derived from multiple sources was triangulated to ensure that the data were not contradictory, and therefore were more likely to describe the investigated program accurately (Lincoln & Guba, 1985; Miles & Huberman, 1984). As recommended by Tobin and Fraser (1998), qualitative methods and quantitative measures were combined to provide insight into the integrated milieu and evaluation of the effects of constructivist pedagogy. Aldridge, Fraser, and Huang (1999) and Tobin and Fraser (1998) combined research methods at different 'grain sizes' to show how individual students and the teacher can be investigated at the class, school, or system level. Aldridge, Fraser, Taylor, and Chen (2000) suggested that the CLES is a useful tool for examining the transformation of teaching and learning practices in Taiwan and Australia in accordance with a constructivist perspective. By combining qualitative and quantitative methods, they determined not only that the learning environments were different in the two countries, but also some explanations of why they were different.

We investigated teacher and student perceptions of the learning environments in terms of the degree of constructivist practice. If the university instructors successfully modelled a constructivist approach, then the question of interest was whether or not the participants' teaching style mirrored that practice (indicating transfer from the university experience into school classrooms). Similar to work reported by Taylor, Fraser, and White (1994), the instructors adopted the role of teacher-researcher in the university classroom, which included summer field trips and online lessons. Qualitative results derived from the instructors' extensive ISLE program experience and in-depth teacher case studies identified four assertions which constitute emergent theory that has a sound empirical base (as reported later in this article).

Research Questions

The study was based primarily on quantitative data derived from the modified version forms of the CLES. Additional qualitative data were derived from reflective field journals, teacher interviews, and classroom observations. The current article reports results based on the use of the CLES and supporting qualitative data from teacher case studies in addressing the following research questions:

1. Did the program instructors foster a constructivist learning environment in the university classroom as perceived by the ISLE teachers?
2. Did the ISLE teachers foster a more constructivist learning environment compared to other classrooms at their same school as perceived by their public/private school students?
3. What assertions can be made about to the implementation of constructivist practice by instructors at the university level?
4. What assertions can be made about the implementation of constructivist practice by teachers at the public/private school level?

Program Design and Delivery

The scales of the CLES provided a critical scaffold for program development. In fact, the research methods employed were integrated into the overall design in an overt manner in order to model the evaluation and assessment of teaching and learning based on the constructivist paradigm. The program was carefully organized to promote constructivist pedagogy within the university classroom and field-based segments. Although each unit focused on a different science topic, the lesson structure followed a similar format for each. Five units in the summer session were comprised of an oncampus meeting, an onsite field experience, and an online component. In the academic year, the same five units included three oncampus laboratory-based classes, one formal seminar presentation, and one reflective online exercise. A sequence of experience, reflection, generalization, and application provided the framework for each lesson within the units.

The resulting 'tapestry' is most clearly evidenced in the electronic discussion board structure. In the summer, each unit developed a knowledge and understanding of a particular aspect of constructivism (as reflected in one of the CLES scales) across content areas, as shown in Figure 1. Throughout the academic year, each unit was discussed in terms of all CLES scales (i.e., fostering Personal Relevance, involving Critical Voice, engaging Student Negotiation, and managing Shared Control), as shown in Figure 2.

The summer and academic-year units were linked through culminating Teaching Roadmap presentations. Participants teamed to 'weave' the complementary experiences into multidisciplinary units addressing defined content. Each of these unit projects included five scripted slides to define the lesson context (level and skills), pedagogical framework (prerequisite and advanced knowledge), logistical framework (5Es: Engage, Explore, Explain, Elaborate, and Evaluate), classroom application details (objective and questioning hierarchy), and topic connections (science disciplines and subject areas). This critical design feature helped to transform the facts gathered from an independent, subject-based division into an integrated, concept-based continuation. Teacher understanding developed as the focus flowed from general observations to specific details and back into the context of the original – but now sharper – 'big picture'.

Use of CLES with Teacher and Student Samples

In the present study, we used a shortened version of the CLES. Concurrent with the work of Nix (2002), Johnson and McClure (2004) used information from factor analysis and internal consistency analysis, as well as the description of scales, to eliminate or rewrite items that were redundant or confusing. This revised and more parsimonious form of the CLES contains only 20 items, with four items in each of the original five scales. In our study, however, we omitted the Uncertainty of Science scale (because of its limited direct relevance to our study), to form the 16-item four-scale version shown in the Appendix. Strategically integrated and deliberately implemented, the CLES defines pedagogical aspects of a classroom learning environment over which the teacher has control and it can be used for evaluating a multidimensional program (Nix, Ledbetter, & Fraser, 2004).

Just as Allen and Fraser (2007) used the same questionnaire to assess multiple views of the learning environment, we evolved two modified versions of the CLES in our study. The adult form (CLES-A) was used to assess teacher perceptions of the university classroom. The comparative student form (CLES-CS) asked secondary students to compare the degree to which they felt that the principles of constructivism had been implemented in the class taught by their ISLE teacher relative to classes taught by other teachers in their school. In the comparative student form, two separate response blocks for each item (THIS and OTHER) are presented in side-by-side columns.

Of primary methodological importance, the CLES provides a valid and reliable instrument for the assessment of how teachers' and students' perceptions of constructivist classroom learning environments change with first-hand experience (see the past studies reviewed above). The established validity of the CLES was important when selecting it to answer the overarching research question of whether or not a teacher's participation in the ISLE program would lead to the teachers' implementation of constructivist learning environments in their respective school classrooms.

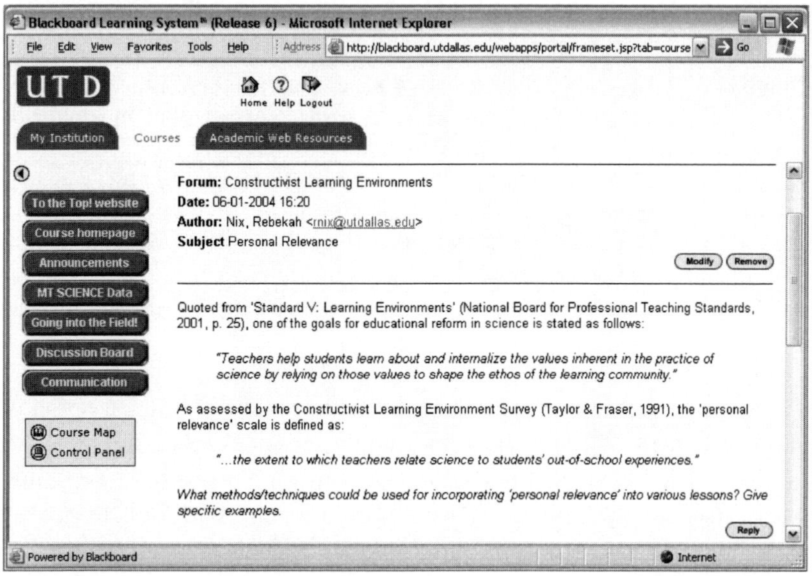

Figure 1. Sample Discussion Board for the Summer Field-Based Session (one CLES scale each unit).

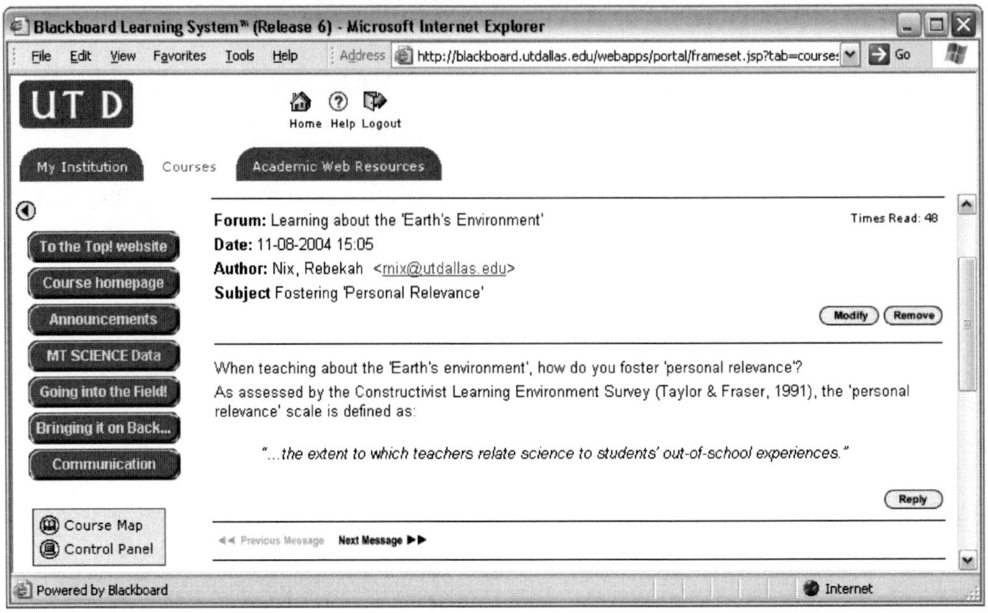

Figure 2. Sample Discussion Board for the Academic Year Laboratory Session (one CLES scales for each unit).

Table 1. Constructivist Learning Environment Survey (CLES) Scales Matched to Learning Environment Goals for Educational Reform in Science

CLES Scale	Science Learning Environment Standard Statement
Personal Relevance	"Teachers help students learn about and internalize the values inherent in the practice of science by relying on those values to shape the ethos of the learning community."
Critical Voice	"...teachers recognize that the emotional response of some students to a lively, argumentative, inquiry-based classroom might be never to venture an opinion or idea, thereby avoiding the risk of public failure."
Shared Control	"Accomplished science teachers deliberately foster settings in which students play active roles as science investigators in a mutually supportive learning community."
Student Negotiation	"They (the teachers) foster a sense of community by encouraging student interactions that show concern for others, by dealing constructively with socially inappropriate behavior, and by appreciating and using humor."

Quoted from Standard V: Learning Environments (National Board for Professional Teaching Standards, 2001, p. 25).

Of primary theoretical importance, the four scales contained in our version of the CLES directly support the goals of educational reform in science described in the Adolescence and Young Adulthood/Science Standards (National Board for Professional Teaching Standards, 2001). Table 1 matches the four CLES scales used in our study to the Science Learning Environments Standard stated as the primary goals for educational reform in the United States.

The adult form of the CLES was administered to all 17 teacher participants in the ISLE program in order to evaluate the instructors' teaching-learning environment for both the summer field session and university classroom lessons. Descriptive statistics (means and standard deviations) were used to depict the overall perceptions of the learning environment held by teachers during their coursework.

Toward the end of the school year, the teachers administered the comparative student form of the CLES (CLES-CS, see the Appendix) to their respective middle-school science students (N = 845). This assessed the degree to which the principles of constructivism were perceived as being implemented in the classes taught by their current teachers who had been program participants (THIS class) relative to classes taught by other teachers in their school (OTHER class). MANOVA for repeated measures was used to ascertain the statistical significance of differences between THIS and OTHER class. Also, effect sizes (Thompson, 1998) were calculated to portray the magnitude, and therefore the educational importance, of differences between THIS and OTHER class. The effect size was calculated by dividing the difference between means by the pooled standard deviation, and it expresses the difference in standard deviation units.

Teachers were asked to emphasize to their school students that there were no right or wrong answers and that students' opinions were what was wanted. Students were encouraged to think about how well each statement describes what the classes are like for them personally, comparing how often each practice occurred in THIS particular science class to OTHER classes that they were taking. Students were directed to read each statement and think about lessons that they had been taught, indicating the best response for the teaching in OTHER classes in the left column. Then they were encouraged to read the statement again, and think about lessons that they had been taught, indicating the best response for the teaching in THIS class in the right column.

RESULTS

Teacher Data

The average item mean score for each CLES-A scale for the 17 teacher participants is depicted graphically in Figure 3. The response alternatives for each item are Almost Never (scored 1), Seldom, Sometimes, Often, and Almost Always (scored 5). Results are shown separately for teachers' perceptions of the intensive one-month summer unit and for the extended program of instruction over a whole academic year. Not surprisingly, teachers' CLES scores are somewhat higher for the extended program over a whole academic year than for the intensive summer unit.

Figure 3 suggests that the practices encompassed by all four scales of the CLES were perceived to occur with an overall frequency of between Often (average item mean of 4) and Almost Always (average item mean of 5) for the summer session and for the extended program. Therefore, in answer to Research Question 1, the data suggest that the program was effective in terms of the degree of implementation of constructivist teaching in the university course, as perceived by the teachers.

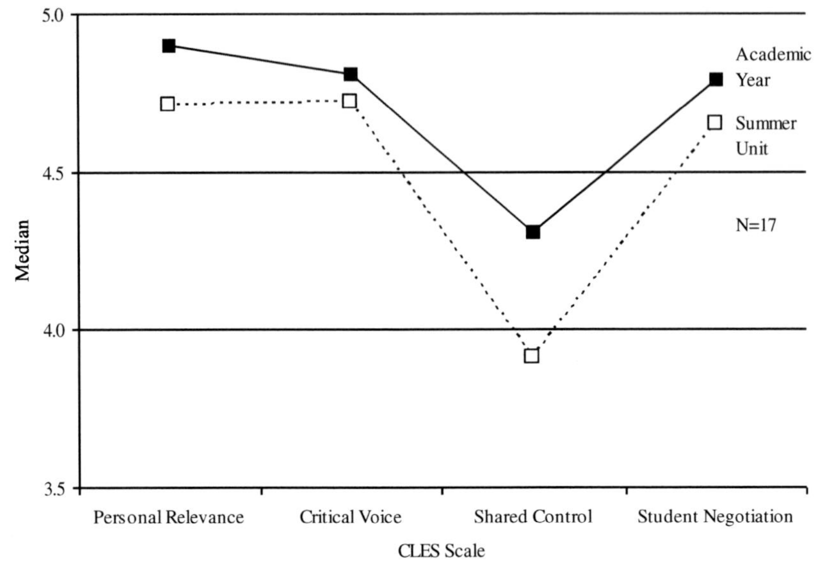

Figure 3. Teacher Perceptions on the Adult Form of the Constructivist Learning Environment Survey (CLES2-A) for the Intensive Summer Unit and for the Extended Program Over an Academic Year.

Crossvalidation of CLES

Because the shortened version of the CLES containing only four items per scale had been validated previously in only one study (Johnson & McClure, 2004), it was important to undertake crossvalidation with our sample. Therefore, for the responses from our sample of 845 students, principal axis factoring with oblique rotation and Kaiser normalization was conducted separately for the 16 items of the CLES-CS for data for THIS class and OTHER class. As recommended by Field (2005) and Stevens (1992), the criteria that we adopted for retention of any item were that its factor loading must be at least 0.40 on its own scale and less than 0.40 on each of the other three scales.

Table 2 shows that, for both for THIS class and OTHER class data, the *a priori* four-factor structure was replicated perfectly and that every item was retained (as its factor loading was greater than 0.40 on its own scale and less than 0.40 on the other three scales). The bottom of Table 2 shows that the proportion of variance accounted for by different CLES scales ranged from 6.77% to 16.19% (with a total of 44.17%) for THIS class and from 6.44% to 15.15% (with a total of 42.27%) for OTHER class. Overall, the results in Table 2 support the factorial validity of the four-scale 16-item version of the CLES.

In order to estimate the internal consistency reliability of each CLES scale, the Cronbach alpha coefficient was calculated separately for THIS and OTHER data. The bottom of Table 2 shows that alpha coefficients for different CLES scales ranged from 0.60 to 0.97 for THIS class and from 0.62 to 0.77 for OTHER class. These values represent satisfactory reliability for short scales containing only four items each.

Table 2. Factor Loadings, % Variance and Alpha Reliability Coefficient for Each Scale of Constructivist Learning Environment Survey (CLES-CS) for THIS and OTHER Class.

	Factor loadings							
	Personal Relevance		Critical Voice		Shared Control			Student Negotiation
Item	THIS	OTHER	THIS	OTHER	THIS	OTHER	THIS	OTHER
PR1	0.63	0.63						
PR2	0.59	0.62						
PR3	0.50	0.50						
PR4	0.63	0.62						
CV1			0.62	0.58				
CV2			0.58	0.59				
CV3			0.49	0.47				
CV4			0.41	0.42				
SC1					0.60	0.40		
SC2					0.61	0.56		
SC3					0.63	0.59		
SC4					0.48	0.43		
SN1							0.65	0.58
SN2							0.69	0.67
SN3							0.75	0.74
SN4							0.59	0.61
% variance	10.80	10.59	6.77	6.44	10.41	10.08	16.19	15.15
Alpha rel.	0.67	0.68	0.62	0.60	0.69	0.66	0.97	0.77

Table 3. Average Item Mean, Average Item Standard Deviation, Effect Size, and MANOVA with Repeated Measures for Differences between Students' Perceptions of ISLE and Other Teachers in the Same School (THIS and OTHER) on the Revised Comparative Student Form of the Constructivist Learning Environment Survey (CLES-CS).

Scale	Items	Average item mean		Average item standard deviation	Difference		
		THIS	OTHER	THIS	OTHER	Effect Size	F
Personal Relevance	4	3.79	3.00	0.83	0.90	0.91	4.92*
Critical Voice	4	3.79	3.60	0.91	0.92	0.21	2.77*
Shared Control	4	3.08	2.90	0.99	0.96	0.18	2.60*
Student Negotiation	4	3.19	3.06	1.10	1.06	0.12	2.09*

* $p < 0.001$.

$N = 845$ students in 45 classes.

THIS refers to the ISLE science teacher's current class; OTHER refers to classes taught by other teachers in the same school.

Item responses are Almost Never (scored 1), Seldom, Sometimes, Often, and Almost Always (scored 5).

Program Effectiveness

Using the CLES-CS, the effectiveness of the ISLE program (which incorporates ICT) was evaluated partially in terms of the degree to which teachers implemented constructivist teaching in their secondary school classrooms, as perceived by the 845 students of the teachers who had experienced ISLE. In order to determine the statistical significance of differences in students' perceptions of the emphasis on constructivist approaches present in their current science teacher's class (THIS) as compared with other teachers' classes (OTHER), we used MANOVA for repeated measures. Because the multivariate test using Wilks' lambda criterion yielded significant results for the set of five CLES scales as a whole, the univariate ANOVA for repeated measures was interpreted for each individual CLES scale. We also calculated the effect size (the difference in means divided by the pooled standard deviation) in order to express the magnitude of differences in standard deviation units (see Table 3).

Table 3 shows that differences between the classroom environments of THIS and OTHER were statistically significant ($p<0.001$) for all CLES scales, indicating that students perceived that participant teachers' classrooms to be more constructivist than other teachers' classrooms in the same school. Effect sizes ranged from around one-tenth of a standard deviation for Student Negotiation to nearly one standard deviation for Personal Relevance. As suggested in previous studies, the smaller effect sizes could suggest areas over which the school administration appear to have strict control. By the same token, the larger effect size suggests that the program could have had an educationally-important effect in improving the indicators over which teachers evidently feel that they have some control (especially Personal Relevance). In answer to Research Question 2, the data suggest that the program was effective in terms of the degree of implementation of constructivist teaching approaches in the ISLE teachers' public/private school classrooms, as perceived by their students.

University-Level Assertion

This section focuses on a school-level assertion that emerged from our qualitative data collection described previously (see Research Question 3). Dr Ledbetter, Professor in Science Education, teamed with Dr Nix, Lecturer in Teacher Development, to design and deliver this program. As suggested by Davis (2004), roles were focused to provide effective support for the diverse needs of participants most efficiently. These new duties required constant collaboration at a variety of levels. The challenge was met by leveraging individual strengths in teaching and research. For example, in terms of research interests, Ledbetter focused on hands-on, inquiry-based science and student misconceptions, while Nix supported technology-rich learning environments and professional development programs. Teaching areas were balanced with respect to field ecology and science education (Ledbetter) and educational technology and instructional design (Nix).

Although the instructors had worked together intuitively to develop and deliver courses and programs for over a decade, this was the first time when they were guided by professional goals (in contrast to program goals) to blend their efforts formally. Not only was this rewarding task easier, but the impact was clearer and greater. Therefore, to answer Research Question 3, our assertion with respect to the implementation of constructivist practice by instructors at the university level is: *An interdisciplinary team approach to program design and delivery provides a critical perspective.*

School-Level Assertions

To answer Question 4, in-depth teacher case studies were undertaken with respect to each CLES scale. Melissa had taught 6th grade science at an affluent, private, preparatory school for four years. As a result of the ISLE program, she claimed that "I have gained the critical professional efficacy needed to enliven my classroom with my personal passion for science education. Understanding the true nature of constructivist theory helped me to make real connections between learning and teaching that made sense for me and my students" (Carpenter & Nix, 2005, p. 1).

Personal Relevance is a vital part of science instruction. As students begin to see how science can affect, influence, and direct other areas of the curriculum, interdisciplinary studies become more interesting and challenging. Regardless of what students learn and retain, it makes little difference if they are unable to connect the information to another area of interest. These connections were demonstrated in the Teaching Roadmaps. Following the 5Es design (Engage, Explore, Explain, Elaborate, Evaluate), teacher teams worked to synthesize the content and pedagogy best suited for creating a positive learning environment in their unique classrooms. A typical Roadmap relates a topic in many different directions. These lesson directives helped Melissa to see the 'big picture' and to use inquiry-based science in her middle-school classroom. Her interest now lies in this new direction of inquiry-based science for the benefit of both the teacher and the students by fostering an interest that is sometimes insatiable.

Melissa's identity is synonymous with personal relevance. Norton (2000 cited in Hawkins, 2004) points out that identity is "how a person understands his or her relationship to the world, how that relationship is constructed across time and space, and how the person understands possibilities for the future" (p. 5). As a teacher, Melissa wants her students to understand the possibilities for the future. It follows that our first assertion with respect to the implementation of constructivist practice by teachers at the public/private school level is: *Teacher efficacy must be founded on a solid base of content knowledge.*

Gwen had taught Grades K–6 science at a low-income, minority, inner-city, private school for 10 years. She says that, "in terms of developing the pupils in my own classroom, this unique opportunity provided an arsenal of new and different methods for reaching out to each individual effectively. The instructors modeled techniques to help us to see how the connections can be made through creative repetition and alternative evaluation. The end result was more meaningful learning (for us) that could be directly transferred to our students" (Satterfield-Barjon & Nix, 2005, p. 1).

Critical Voice was promoted through exercises that exposed teachers to the opportunities and encouraged them to hypothesize about the possibilities for their projects. For example, a trip to a local aquarium provided ideas about how to use classroom aquariums and terrariums. Gwen explained that, with a greater understanding of the importance of monitoring ammonia levels in their tanks, students found that the nitrogen cycle became much more relevant and understandable and that her tanks became beneficial learning tools.

Gwen saw Shared Control demonstrated as participants learned more about scientific methodology and focused their individual topics within the group perspective. She noted that the Teaching Roadmap placed less emphasis on student competition and more on cooperation and participation. She was made more aware of how strategic grouping impacts the learning experience and can lead to active and energetic classrooms which, in her opinion, can be a sign of healthy learning.

Student Negotiation was fostered through professional networks initiated and developed throughout the program. Teachers discussed their project ideas and ongoing investigations with scientific experts who were conducting relevant research, informal science education specialists offering curriculum support, community leaders promoting local agendas and, most importantly, with each other as peers and colleagues with a shared goal and unique vision. The interactive professional seminars served as yet another way of moving educators upward. Gwen was encouraged to know that leaders in various fields did not presume the 'I've got mine, now you get yours' attitude.

Theory does not produce change; people do. Having witnessed the benefits and successes of hands-on learning, Gwen's administrators became more open to change. They firmly support, praise, and encourage 'doing science' in all of the school's classrooms when students proclaim that science is their favorite subject! Therefore, our second school-level assertion is: *A teacher's intimate and practical understanding of a subject is prerequisite for successful incorporation of new pedagogical skills.*

Both 6th grade science teachers in a large, suburban district, Ernie had taught in elementary school for eight years and Chris had taught in middle school for two years. On reflection, they realized that "most teachers have access to these same tools through workshops or inservice training; however, the volume of new material is not always easily integrated into actual classroom lessons. The positive outcomes of this program can be attributed to both the long-term nature of the training and the support of the local network. This opportunity to learn in new ways helped us to develop the skills needed to teach in different ways. Sharing ideas and experiences helped us all to discover broader options for facing the similar challenges that we commonly face in our nearby communities and in the field of science education" (Huben, Long, & Nix, 2005, p. 1).

The information revolution has provided overwhelming opportunities and options for virtually anyone who is interested (Nix, 2000). New learning environments are either being created or are presently evolving to supply the demands of local business and the global society. Time and space no longer limit the possibilities for lifelong learning but, compared with the rapid pace of everyday operations fuelled by the information explosion, advances in science education are slow, at best (Kuhn, 1970). Rapidly-changing demographics are evolving north Texas into a less homogenous, urban society. A diversified population requires different types of intervention strategies and a variety of instructional styles at all levels.

According to Ernie and Chris, the 'glue' that held the program together was the online course shell. It facilitated the development of the Teaching Roadmaps because teachers could refer back to those experiences as connections were made across units. Teachers said that, when they got busy with their new school classes, it was especially helpful to have an electronic reference list of pedagogically-sound and scientifically-sound websites. Because these teachers had learned with them, it was easy for them to incorporate them into their own lessons.

The same held true for the electronic data-collection devices. Most participants found that their students also enjoyed using the probes, commenting that the 'high-tech feel' helped with engagement. The ability to take more accurate readings in less time allowed students to ask more questions about the procedure and to design and conduct variations to the experiments on their own (touching on all CLES scales). Borrowing the associated laptop computers freed teachers of district restrictions and extended their reach beyond the physical classroom. Not surprisingly, the availability and expense of materials posed a major issue in most of the schools represented by participants. An array of activity books and curriculum guides provided low-cost, hands-on activities that involved the use of common household items. Teachers said that these are just as effective and that students do not 'feel deprived' because they were using a plastic cup rather than a test tube. In all of the activities performed in the program and repeated in the teachers' classrooms, participants reported high student enthusiasm and active engagement in scientific discovery. 'Take-away' models also could be used to demonstrate activities and, ideally, to serve as samples for school classroom sets.

If it takes a village to raise a child, then the same thing can be said about teaching a child. Informal science educators, scientific researchers, university staff, and district staff and administrators teamed to realize the instructional objectives. Participants also formed a peer network to support each other in implanting effective lessons. Just as in the classroom, the program design balanced individual, team, and group activities in order to deepen the collaboration and strengthen the existing and emerging networks. Besides providing indepth content from specialists, this aspect served to initiate associations that support continuous learning beyond program duration. Familiarization with local environments and informal science education centers helped to develop a 'sense of place' that teachers can use on field trips or when encouraging students to visit.

As lifelong learners, teachers know how to 'fish'; this program simply provided a structure to help them 'choose the bait'. "Teachers need time, assistance, examples, and the appropriate opportunity to try things out and make mistakes, before we can remotely expect appropriate, much less seamless, integration of the technology into day to day teaching" (Kessell, 1997, ¶ 44). Therefore, our final school-level assertion is: *A working knowledge of and the availability of new tools and current resources, along with an active peer network, ultimately determine a teacher's facility for enhancing their students' learning environment.*

CONCLUSIONS

Learning environments research has a broad range of applicability for today's diverse educational issues. The study reported in this chapter provides another example of the use of learning environment variables in educational program evaluation. Modified and shortened versions of the CLES were found to be valid, economical and useful in our study. Quantitative data suggest that, in terms of the scales of the CLES, instructors were successful in fostering a constructivist learning environment in the university classroom as perceived by the teachers, and participating teachers were successful in fostering a more constructivist learning environments compared to other classrooms at their same school as perceived by their public/private school students. Although our previous evaluation of the implementation of the ISLE model within a summer short course also attested to the model's success (Nix et al. 2005a), it is noteworthy that the effect sizes were considerably larger in the current evaluation of the implementation of the model over a three-semester time period.

Consistent with previous studies, the ISLE model offered a broad context for enculturation of the constructivist paradigm. Because of the influence of numerous school-level factors, this sort of pedagogical change is difficult to realize in individual classrooms (Milne & Taylor, 2000). In our study, qualitative data led to four main assertions with respect to the implementation of constructivist teaching-learning practices in universities and public/private schools:

1. An interdisciplinary team approach to program design and delivery provides a critical perspective.
2. Teacher efficacy must be founded on a solid base of content knowledge.
3. A teacher's intimate and practical understanding of a subject is prerequisite for successful incorporation of new pedagogical skills.
4. A working knowledge of and the availability of new tools and current resources, along with an active peer network, ultimately determine a teacher's facility to enhance their students' learning environment.

Based on the ISLE model and the use of ICT, this year-long program merged theory and practice with science and education to create a rich new learning environment. Although "achievement is not the only important outcome of schooling" (Shumway, Saunders, Stewardson, & Reeve, 2001, ¶ 4), it is important to remember that the current body of research shows that positive classroom environments have been positively associated with student attitudes to science and student cognitive achievement in science (Fraser, 1991, 1994, 2007; Fraser, Walberg, Welch, & Hattie, 1987; Haertel et al., 1981; McRobbie & Fraser, 1993).

There were some specific limitations with our study. First, the intervention was restricted to a small sample of practicing science teachers located geographically in the surrounding Dallas metropolitan area. Potentially reducing generalization, this unique cohort was recruited through a federally-funded grant. In addition, the case-study teachers were graduate degree candidates at the institution. Finally, the student body represented in the school classrooms was wide-ranging in demographics and background.

Our study provides evidence-based research that the ICT-enhanced teacher education program achieved one of its main objectives, namely, enhancing learning environments, which is linked directly to the other two objectives of increasing content knowledge and improving pedagogical expertise. Constructivist techniques fostered participants' respect and rapport for modelling a positive climate for learning, equity, and excellence. Long-term modelling of effective science teaching with field-tested resources enabled teachers to cultivate positive learning environments in their school classrooms.

REFERENCES

Ajzen, I. (1985). From intentions to actions: A theory of planned behavior. In J. Kuhl & J. Beckman (Eds.), *Action control: From cognition to behavior* (11-39). New York: Springer Verlag.

Aldridge, J. M., Fraser, B. J. & Huang, T. C. I. (1999). Investigating classroom environment in Taiwan and Australia with multiple research methods. *Journal of Educational Research, 93*, 48-63.

Aldridge, J. M., Fraser, B. J. & Sebela, M. P. (2004). Using teacher action research to promote constructivist learning environments in South Africa. *South African Journal of Education, 24*, 245-253.

Aldridge, J. M., Fraser, B. J., Taylor, P. C. & Chen, C. C. (2000). Constructivist learning environments in a cross-national study in Taiwan and Australia. *International Journal of Science Education, 22*, 37-55.

Allen, D. & Fraser, B. J. (2007). Parent and student perceptions of classroom learning environment and its association with student outcomes. *Learning Environments Research, 10*, 67-82.

Beck, J., Czerniak, C. M. & Lumpe, A. T. (2000). An exploratory study of teachers' beliefs regarding the implementation of constructivism in their classroom. *Journal of Science Teacher Education, 11*, 323-343.

Cannon, J.R. (1995). Further validation of the Constructivist Learning Environment Survey: Its use in the elementary science methods course. *Journal of Elementary Science Education, 7(1)*, 47-62.

Carpenter, M. & Nix, R. K. (2005, November). *New perspectives: Inspiring students to sense and see 'science' everyday*. Paper presented at the annual meeting of the School Science and Mathematics Association, Fort Worth, TX.

Cho, J. I., Yager, R. E., Park, D. Y. & Seo, H. A. (1997). Changes in high school teachers' constructivist philosophies. *School Science and Mathematics,* 97, 400-405.

Committee on Undergraduate Science Education. (1997*). Science teaching reconsidered: A handbook*. Washington, DC: National Academy Press.

Davis, J. R. (2004). Interdisciplinary teaching: Team-taught courses challenge the dominant paradigm. *Thriving in Academe, 22*(2), 5-8.

Dryden, M. & Fraser, B. J. (1998, April). *The impact of systemic reform efforts on instruction in high school classes*. Paper presented at the annual meeting of the American Educational Research Association, San Diego, CA.

Field, A. (2005). *Discovering statistics using SPSS* (2nd ed.). London: Sage Publications.

Fisher, D. L. & Khine, M. S. (Eds.). (2006). *Contemporary approaches to research on learning environments: Worldviews*. Singapore: World Scientific.

Foxon, M. (1993). A process approach to the transfer of training. Part 1: The impact of motivation and supervisor support on transfer maintenance. *Australian Journal of Educational Technology, 9*, 130-143.

Fraser, B. J. (1979). Evaluation of a science-based curriculum. In H.J. Walberg (Ed.), *Educational environments and effects: Evaluation, policy, and productivity* (218-234). Berkeley, CA: McCutchan.

Fraser, B. J. (1991). Two decades of classroom environment research. In B.J. Fraser & H.J. Walberg (Eds.), *Educational environments: Evaluation, antecedents and consequences* (3-27). Oxford: Pergamon Press.

Fraser, B. J. (1994). Research on classroom and school climate. In D.L. Gabel (Ed.), *Handbook of research on science teaching and learning* (493-541). New York: Macmillan.

Fraser, B. J. (1998a). Classroom environment instruments: Development, validity and applications. *Learning Environments Research, 1*, 7-33.

Fraser, B. J. (1998b). Science learning environments: Assessment, effects and determinants. In B. J. Fraser & K. G. Tobin (Eds.), *International handbook of science education* (527-564). Dordrecht, The Netherlands: Kluwer Academic Publishers.

Fraser, B. J. (2001). Twenty thousand hours. *Learning Environments Research, 4*, 1-5.

Fraser, B. J. (2007). Classroom learning environments. In S.K. Abell & N.G. Lederman (Eds.), *Handbook of research on science education* (103-124). Mahwah, NJ: Lawrence Erlbaum.

Fraser, B. J. & Fisher, D. L. (1983). Student achievement as a function of person-environment fit: A regression surface analysis. *British Journal of Educational Psychology, 53*, 89-99.

Fraser, B. J. & Walberg, H. J. (Eds.). (1991). Educational environments: Evaluation, antecedents and consequences. London: Pergamon.

Fraser, B. J., Walberg, H. J., Welch, W. W. & Hattie, J. A. (1987). Syntheses of educational productivity research. *International Journal of Educational Research, 11*, 145-252.

Geelan, D. R. (1997). Weaving narrative nets to capture school science classrooms. *Research in Science Education, 27*, 553-563.

Goh, S. C. & Khine, M. S. (Eds.). (2002). Studies in educational learning environments: An international perspective. Singapore: World Scientific.

Goh, S. C., Young, D. J. & Fraser, B. J. (1995). Psychosocial climate and student outcomes in elementary mathematics classrooms: A multilevel analysis. *Journal of Experimental Education, 64*, 29-40.

Haertel, G. D., Walberg, H. J. & Haertel, E. H. (1981). Socio-psychological environments and learning: A quantitative synthesis. *British Educational Research Journal, 7*, 27-36.

Harwell, S. H., Gunter, S., Montgomery, S., Shelton, C. & West, D. (2001). Technology integration and the classroom learning environment: Research *for* action. *Learning Environments Research, 4,* 259-286.

Hawkins, M. R. (2004). Researching English language and literacy development in schools. *Educational Researcher, 33(3),* 14-25.

Huben, E. W., Long, C. S. & Nix, R. K. (2005, November). *Let's go... Getting your gear in gear to teach science, today*. Paper presented at the annual meeting of the School Science and Mathematics Association, Fort Worth, TX.

Johnson, B. & McClure, R. (2004). Validity and reliability of a shortened, revised version of the Constructivist Learning Environment Survey (CLES). *Learning Environments Research, 7*, 65-80.

Kessell, S. R. (1997, June). *Staff development in computing and information systems literacy: Computing as a tool, not an end in itself.* Paper presented at the Gender and Science and Technology Association-International Organisation for Science and Technology Education Australasian Joint Regional Conference, Curtin University of Technology, Perth, Western Australia.

Khine, M. S. & Fisher, D. L. (Eds.). (2003). *Technology-rich learning environments: A future perspective*. Singapore: World Scientific.

Khoo, H. S. & Fraser, B. J. (2008). Using classroom psychosocial environment in the evaluation of adult computer application courses in Singapore. *Technology, Pedagogy and Education, 17*, 53-67.

Kim, H. B., Fisher, D. L. & Fraser, B. J. (1999). Assessment and investigation of constructivist science learning environments in Korea. *Research in Science and Technological Education, 17*, 239-249.

Kuhn, T. (1970). *The structure of scientific revolutions*. Chicago: The University of Chicago Press.

Lightburn, M. E. & Fraser, B. J. (2007). Classroom environment and student outcomes among students using anthropometry activities in high school science. *Research in Science and Technological Education, 25*, 153-166.

Lincoln, Y. S. & Guba, E.G. (1985). *Naturalistic inquiry*. Beverly Hills, CA: Sage Publications.

Maor, D. & Fraser, B. J. (1996). Use of classroom environment perceptions in evaluating inquiry-based computer-assisted learning. *International Journal of Science Education, 18*, 401-421.

Martin-Dunlop, C. & Fraser, B.J. (2008). Learning environment and attitudes associated with an innovative course designed for prospective elementary teachers. *International Journal of Science and Mathematics Education, 6*, 163-190.

McRobbie, C. J. & Fraser, B. J. (1993). Associations between student outcomes and psychosocial science environments. *Journal of Educational Research, 87*, 78-85.

Miles, M. B. & Huberman, A. M. (1984). *Qualitative data analysis: A sourcebook of new methods*. Beverly Hills, CA: Sage Publications.

Milne, C. & Taylor, P. (2000, April). *"Facts are what you teach in science!" Teacher beliefs and the culture of school science*. Paper presented at the annual meeting of the National Association for Research in Science Teaching, New Orleans, LA.

Moscovici, H. (1999, March). *A theoretical model for shifting toward inquiry science experiences: How does it work?* Paper presented at the annual meeting of the National Association for Research in Science Teaching, Boston, MA.

National Board for Professional Teaching Standards. (2001). *Adolescence and young adulthood/science standards* (2nd printing). Washington, DC: Author.

Nix, R. K. (2000). Science-related virtual field trips on the world wide web. *The Texas Science Teacher, 29(1)*, 18-23.

Nix, R. K. (2002). *Virtual field trips: Using information technology to create an integrated science learning environment*. Unpublished doctoral thesis, Curtin University of Technology, Perth, Western Australia.

Nix, R. K., Fraser, B. J. & Ledbetter, C.E. (2005a). Evaluating an integrated science learning environment using the Constructivist Learning Environment Survey (CLES). *Learning Environments Research, 8*, 109-133.

Nix, R. K., Fraser, B. J. & Ledbetter, C. E. (2005b, August). *Using science student perceptions of the classroom learning environment to evaluate a professional development program*. Paper presented at the International Conference on Science, Mathematics, and Technology Education, Victoria, BC, Canada.

Nix, R. K. & Ledbetter, C. E. (2005, November). *Cultivating an integrated science learning environment (ISLE) for middle school teachers*. Paper presented at the annual meeting of the School Science and Mathematics Association, Fort Worth, TX.

Nix, R. K., Ledbetter, C. E. & Fraser, B. J. (2001, December). *Evaluation of an integrated science learning environment that bridges university field classes and field trips*. Paper presented at the annual meeting of the Australian Association for Research in Education, Fremantle, Australia. [ISSN 1324-9320.]

Nix, R. K., Ledbetter, C. E. & Fraser, B. J. (2004, April). Designing, delivering, and evaluating a field-based science course for teachers using the Constructivist Learning Environment Survey (CLES). Paper presented at the annual meeting of the National Association for Research in Science Teaching, Vancouver, BC, Canada.

Ogbuehi, P. I. & Fraser, B. J. (2007). Learning environment, attitudes and conceptual development associated with innovative strategies in middle-school mathematics. *Learning Environments Research, 10*, 101-114.

Oh, P. S. & Yager, R. E. (2004). Development of constructivist science classrooms and changes in student attitudes toward science learning. *Science Education Journal, 15*, 105-113.

Peiro, M. M. & Fraser, B. J. (in press). Assessment and investigation of science learning environments in the early childhood grades. In M. Ortiz & C. Rubio (Eds.), *Educational evaluation: 21st century issues and challenges*. New York: Nova Scientific Publishers.

Rillero, P. (1993, April). *The revolution of enlightenment: A historical case study of significant educational change through teacher education*. Paper presented at the annual meeting of the National Association of Research in Science Teaching, Atlanta, GA.

Satterfield-Barjon, G. F. & Nix, R. K. (2005, November). *Professional development-to-pupil development: Building stronger links by 'doing science'*. Paper presented at the annual meeting of the School Science and Mathematics Association, Fort Worth, TX.

Shulman, L. (1990). *Aristotle had it right: On knowledge and pedagogy* (Occasional paper no. 4). East Lansing, MI: The Holmes Group.

Shumway, S., Saunders, W., Stewardson, G. & Reeve, E. (2001). A comparison of classroom interpersonal goal structures and their effect on group problem-solving performance and student attitudes toward their learning environment. *Journal of Industrial Teacher Education, 38*(3). [On-line.] Available: http://scholar.lib.vt.edu/ejournals/JITE/v38n3/shumway.html.

Spinner, H. & Fraser, B. J. (2005). Evaluation of an innovative mathematics program in terms of classroom environment, student attitudes, and conceptual development. *International Journal of Science and Mathematics Education, 3*, 267-293.

Stevens, J. P. (1992). *Applied multivariate statistics for the social sciences* (2nd ed.). Hillsdale, NJ: Erlbaum.

Taylor, P. C. & Fraser, B. J. (1991, April). *Development of an instrument for assessing constructivist learning environments*. Paper presented at the annual meeting of the American Educational Research Association, New Orleans, LA.

Taylor, P. C., Fraser, B. J. & Fisher, D.L . (1997). Monitoring constructivist classroom learning environments. *International Journal of Educational Research, 27*, 293-302.

Taylor, P. C., Fraser, B. J. & White, L. R. (1994, April). *CLES: An instrument for monitoring the development of constructivist learning environments*. Paper presented at the annual meeting of the American Educational Research Association, New Orleans, LA.

Teh, G. & Fraser, B. J. (1994). An evaluation of computer-assisted learning in terms of achievement, attitudes and classroom environment. *Evaluation and Research in Education, 8*, 147-161.

Thompson, B. (1998). Review of 'what if there were no significance tests?' *Educational and Psychological Measurement, 58*, 334-346.

Tobin, K. & Fraser, B. J. (1998). Qualitative and quantitative landscapes of classroom learning environments. In B. J. Fraser & K. G. Tobin (Eds.), *International handbook of science education* (623-640). Dordrecht, The Netherlands: Kluwer.

Tobin, K., Tippins, D. J. & Gallard, A. (1994). Research on instructional strategies for teaching science. In D.L. Gabel (Ed.), *Handbook of research on science teaching and learning* (45-93). New York: Macmillan.

von Glaserfeld, E. (1995). Radical constructivism: A way of knowing and learning. London: The Falmer Press.

Zandvliet, D. B. & Fraser, B. J. (2004). Learning environments in information and communications technology classrooms. *Technology, Pedagogy and Education, 13*, 97-123.

APPENDIX

Items (Grouped by Scale) for the Revised Comparative Student Form of the Constructivist Learning Environment Survey (CLES-CS)

Item	Statement
Personal Relevance	
PR1	I learn about the world inside and outside of school.
PR2	New learning relates to experiences or questions about the world inside and outside of school.
PR3	I learn how my studies are a part of my life inside and outside of school.
PR4	I learn interesting things about the world inside and outside of school.
Critical Voice	
CV9	I feel safe questioning what or how I'm being taught.
CV10	I feel I learn better when I am allowed to question what or how I'm being taught.
CV11	It's OK for me to ask for clarification about activities that are confusing.
CV12	It's acceptable for me to express concern about anything that gets in the way of my learning.
Shared Control	
SC13	I help plan what I am going to learn.
SC14	I help to decide how well I am learning.
SC15	I help to decide which activities work best for me.
SC16	I let the teacher know if I need more/less time to complete an activity.
Student Negotiation	
SN17	I talk with other students about how to solve problems.
SN18	I explain my ideas to other students.
SN19	I ask other students to explain their ideas.
SN20	I am asked by others to explain my ideas.

Adapted from Taylor & Fraser (1991) and Johnson & McClure (2004).

The response alternatives are Almost Never (scored 1), Seldom, Sometimes, Often, and Almost Always (scored 5).

In: Computer-Assisted Teaching: New Developments ISBN: 978-1-60876-855-4
Editors: B. A. Morris, G. M. Ferguson, pp. 117-132

Chapter 5

IMPLICATIONS OF EXPERTISE REVERSAL EFFECT FOR MULTIMEDIA LEARNING

Slava Kalyuga*

University of New South Wales, Sydney, Australia.

ABSTRACT

Expertise reversal effect relates to the interactions between levels of learner prior knowledge and effectiveness of different instructional techniques and procedures. Designs and techniques that are effective with low-knowledge learners can lose their effectiveness and even have negative consequences for more proficient learners. This paper reviews recent empirical findings associated with the expertise reversal effect in multimedia and hypermedia learning environments, their interpretation within a cognitive load framework, and implications for the design of learner-tailored multimedia.

INTRODUCTION

Learner-tailored forms of instruction have for long time been recognized as beneficial for students, however most instructional materials continue to be designed in a uniform way, and usually assume novices as intended learners. Research in adaptive multimedia and hypermedia systems has been focused more on technical issues of tailoring instructional content to learner preferences, interests, choices, history of previous on-line behavior etc. Fundamental cognitive characteristics of learners have not been seriously considered by designers of multimedia instructional materials.

Research on expert-novice differences has demonstrated that learner knowledge base is the most important cognitive characteristic that influences learning. Recent studies within the cognitive load framework have demonstrated that designs and techniques that are effective with low-knowledge individuals can lose their effectiveness and even have negative consequences for more proficient learners (see Kalyuga, 2005; 2006b; 2007 for recent

* Corresponding author: E-mail: s.kalyuga@unsw.edu.au

overviews). The reversal in the relative effectiveness of instructional methods as levels of learner knowledge in a domain change has been called an expertise reversal effect. The major instructional design implication of this effect is the need to adjust dynamically instructional methods and procedures as learners acquire more expertise in a specific domain.

The following sections of the paper briefly describe relevant characteristics of our cognitive architecture that underlie the expertise reversal effect, main empirical findings associated with this effect in multimedia and hypermedia learning environments, their interpretation within cognitive load theory, and implications for the design of learner-tailored multimedia environments.

Role of Expertise in Cognitive Processes

Our cognitive system has several major characteristics that define how we learn and perform in multimedia environments. Firstly, cognitive processes essentially depend on a knowledge base in long-term memory with effectively unlimited capacity and duration. Organized generic knowledge structures (schemas) are used to mentally categorize and represent concepts and procedures, and govern our behavior. Secondly, our cognitive system includes a mechanism that limits the scope of immediate simultaneous changes to the knowledge base. This mechanism is associated with the concept of working memory as a conscious processor of information within our focus of attention. Working memory is severely limited in capacity and duration when dealing with novel information (Cowan, 2001; Miller, 1956). Working memory is believed to have separate limited processing channels for visual and auditory information modalities (Baddeley, 1986). Processing limitations of working memory and associated cognitive load profoundly influence the effectiveness of instruction.

However, the most important factor of our cognitive functioning is how these two systems interact with each other. Knowledge base in long-term memory allow us to effectively reduce limitations of working memory by encapsulating many elements of information into larger, higher-level units that are treated as elements in working memory. Another way to reduce cognitive load is to practice skills until they can operate under automatic rather than controlled processing (Kotovsky, Hayes, & Simon, 1985; Shiffrin & Schneider, 1977). When basic routine procedures occur automatically, the system could avoid an overload and reallocate cognitive resources for higher-level mental processes. When learners do not have relevant knowledge (or if it is not sufficiently automated), they have to deal with many new elements of information that may easily overload working memory. These learners may require considerable external support to build new knowledge structures in a relatively efficient manner. On the other hand, more knowledgeable learners may rely on their available domain-specific long-term memory structures for managing cognitive load.

Knowledge structures in long-term memory also perform an organizing and governing role in complex cognitive processes. In the absence of relevant knowledge or sufficient external instructional guidance, we would use mostly random search processes in attempts to handle the task. If no guidance is provided for dealing with new units of information, the task may cause a cognitive overload. However, if guidance is provided to learners who have sufficient knowledge base for dealing with the same units of information, learners would have to relate and reconcile the related components of the available long-term memory base and external information. Such integration processes may impose an additional cognitive load and reduce resources available for learning new knowledge, thus causing an expertise reversal effect. Presenting knowledgeable learners with detailed external guidance may hinder their learning and performance relative to the levels they could achieve with minimal instructional support. Therefore, as the level of learner expertise in a domain increases, relative effectiveness of learning tasks with different levels of instructional support may reverse.

Although the expertise reversal effect was first predicted within the cognitive load theoretical framework as a form of redundancy effect that could occur when some presented information that was beneficial (and non-redundant) for novice learners became redundant for learners with higher levels of knowledge in a task domain (Kalyuga, Chandler, & Sweller, 1998), the effect was then extended to different presentation modalities and levels of instructional guidance.

SOURCES OF COGNITIVE LOAD

Cognitive load is the demand for working memory resources that are required for achieving goals of a learning task. One type of load, an intrinsic cognitive load, is essential for learning and caused by internal complexity of the task relative to the level of learner expertise. For example, expertise in the language of instruction may influence the level of intrinsic cognitive load, e.g., simple and routine sentences for native speakers may cause significant cognitive overload for second language learners. This type of cognitive load is required for comprehending a situation and results in modified or new knowledge structures in long-term memory. Therefore, it is vital to provide all the necessary resources to accommodate this load without exceeding limits of working memory capacity. For example, to manage intrinsic cognitive load, the learning goal could be divided into a series of sub-goals that require less processing resources, or instructional tasks could be segmented into smaller units.

Alternatively, some of the essential interactions between elements of information could be excluded from consideration in order to artificially reduce structural complexity of the task on initial stages of learning followed by the fully interactive materials later (Pollock, Chandler, & Sweller, 2002). If intrinsic load is at low levels and much cognitive capacity remains unused, it could be increased, for example, by setting more challenging learning goals that require more complex cognitive activities with higher levels of element interactivity.

In contrast to essential, extraneous cognitive load is an irrelevant form of load associated with a waste of cognitive resources due poor presentation designs, inappropriate selection and sequencing of learning tasks, or inadequate instructional support. For example, extraneous cognitive load could be caused by separating related sources of information in space and/or time; duplicating the same information simultaneously in different modalities; or using unguided problem-solving or exploratory activities with novice learners.

The expertise reversal effect is associated with two types of situations that cause extraneous cognitive load: 1) insufficient external guidance does not compensate for limited knowledge of novice learners; 2) expert learners' knowledge base overlaps with provided external guidance thus forcing learners to waste limited resources on co-referring internal and external representations of the same information. It should be noted that the difference between extraneous and intrinsic cognitive load is relative to levels of learner expertise: some components of cognitive load that are essential for novice learners could become extraneous (irrelevant) for relatively more experienced learners, and vice versa.

LEVELS OF LEARNER EXPERTISE AND COGNITIVE LOAD EFFECTS

When different sources of information that require mental integration for understanding are separated in space or time, the process of integration (e.g., visual search-and-match) may substantially increase cognitive load and inhibit learning. Physically integrated or embedded formats were demonstrated to be an effective alternative to "split-source" instructions (split-attention effect; e.g., Mayer & Gallini, 1990; Sweller, Chandler, Tierney, & Cooper, 1990).

Split-attention situations may also be managed by using different modalities, since working memory includes separate subsystems for storing and processing visual and auditory information (Baddeley, 1986). Integration of the verbal auditory and pictorial visual information may not overload working memory if its capacity is effectively expanded by using a dual-mode presentation (modality effect; e.g., Mayer, 1997; Mousavi, Low, & Sweller, 1995). For example, Mayer and Anderson (1991) compared an animation depicting the operation of a bicycle tire pump with simultaneous audio text; the audio text only without an animation; the animation only without audio text; and a control condition with no formal training. According to the post-training test, the first experimental condition produced the best results. Mayer & Anderson (1992) and Mayer & Sims (1994) compared simultaneous and sequential presentations of the related audio and visual information and demonstrated that dual-mode instructions were superior only when presented in the simultaneous form (the temporal contiguity effect).

Cognitive load measures were not used in the above experiments that were generated within the framework of cognitive theory of multimedia learning. Based on cognitive load theory, Tindall-Ford, Chandler, & Sweller (1997) predicted a modality effect based on the idea of extending effective working memory capacity. The prediction was empirically supported by experimental studies in training technical apprentices. Learning from dual-mode instructions (inspecting electrical wiring diagrams and listening to simultaneous auditory explanations) produced better post-test performance than equivalent single-mode presentations (studying wiring diagrams with printed textual explanations). Importantly, measures of cognitive load using subjective rating scales corroborated the cognitive load-based hypothesis: dual-mode instructions imposed less cognitive load as reported by trainees themselves.

However, if different sources of information are intelligible in isolation, elimination rather than visual or auditory integration of a redundant source could be preferable (redundancy effect; e.g., Chandler & Sweller, 1991; Mayer, Heiser, & Lonn, 2001). However, whether information is redundant depends on the level of expertise of the learner: what is essential for novices could be redundant for more knowledgeable learners. When onscreen text is embedded into a diagram or narrated when the diagram is presented, it is not possible to avoid processing the redundant information and integrating it with available knowledge structures in long-term memory. These processes consume cognitive resources that become unavailable for constructing higher levels of knowledge. Eliminating redundant verbal or pictorial information could be beneficial for more experienced learners.

Thus, the effectiveness of different instructional formats may depend on levels of expertise of the intended learners (in accordance with the expertise reversal effect). While novice learners may learn best from textual explanations embedded into a diagram or narrated over the diagram, for more experienced learners, diagram-alone materials could generate higher levels of performance and be easier to process. Textual explanations that are essential for novices may become redundant for experts. For example, according to Kalyuga, Chandler, and Sweller (2000), replacing on-screen textual explanations of procedural steps involved in reading specific charts used to set up cutting machines with corresponding narrated explanations was beneficial for novice learners (thus demonstrating a modality effect). Interactive diagram-only presentations (that provided student-controlled on-screen highlights of lines, numbers and other components of the chart) were the least effective for these inexperienced learners. Following the initial learning phase, all learners went through a series of common intensive training sessions and became more experienced in the task domain. When they were trained again in more advanced types of charts using alternative experimental instructional formats, an interactive diagram-only presentation format became more beneficial than the diagram with narrated explanations. Subjective rating scales indicated relatively lower levels of cognitive load imposed on learners by the diagram-only format.

The pattern of results in this set of experiments is typical for many studies of the expertise reversal effect. For example, the techniques for enhancing levels of instructional guidance that were effective for novice learners (e.g., using worked examples instead of conventional problem solving or exploratory learning) became ineffective and often resulted in negative rather than positive or neutral effects for more knowledgeable learners (Kalyuga, 2006b). Narrated explanations (e.g., audio-taped college lectures) could be supported by different types of visual overhead aids: knowledge maps, topic outlines, or lists of key terms. Lambiotte and Dansereau (1992) demonstrated that students with low prior knowledge benefited most from the knowledge maps and least from the lists of key terms. For more knowledgeable learners, the results reversed with the lists of key terms as the most effective learning aid.

ROLE OF LEARNER EXPERTISE IN PROCESSING VISUAL REPRESENTATIONS

R. Mayer and his associates studied learning from text and graphics using learners with different levels of prior knowledge. The experiments indicated that using graphics usually enhanced learning outcomes for students with low prior knowledge levels, but not those with higher prior knowledge levels. In his major review of the multimedia learning studies, Mayer (2001) called the effect an individual differences principle. It was attributed to the ability of high-knowledge learners to use their knowledge base to compensate for missing instructional guidance. Advantages of multimedia presentations disappeared with increases in learner levels of prior knowledge.

For example, Mayer and Gallini (1990) and Mayer, Steinhoff, Bower, and Mars (1995) demonstrated that spatially integrated instructional formats were more effective for low-knowledge learners rather than for high-knowledge learners. As learners' knowledge in a domain increases, positive learning effects of such formats disappeared. Kalyuga et al. (1998) found that more experienced technical apprentices learned new versions of a familiar class of wiring diagrams more effectively from the diagrams alone rather than from diagrams with integrated detailed explanations. Additional instructional explanations were redundant for these learners and generated extraneous cognitive load (as reported by learners' subjective ratings of cognitive load).

Thus, advantages of pictorial representations may disappear with increases in learner levels of expertise. Lee, Plass, and Homer (2006) investigated two different modes of visual representations in a gas law simulation for middle-school chemistry students. Essential gas characteristics were presented either in symbolic form only (words 'temperature', 'pressure', and 'volume' with corresponding numerical values) or by adding iconic information to the symbolic representations (e.g., burners for temperature, weighs for pressure). While low prior knowledge learners benefited more from added iconic representations than from symbolic formats only, high prior knowledge learners benefited more from symbolic only representations. Iconic representations were redundant for these learners.

Since animations and simulations belong to visual representations in general, cognitively-based principles for designing and using visual representations apply to animations as well (Rieber, 1990). On the other hand, animations are dynamic visualizations that are capable of representing movements and processes, and this is the most important reason for using animations in instruction. Animations could be effective instructional tools for concepts and situations that are difficult to describe and/or illustrate verbally (e.g., abstract concepts and processes, changes over time, trajectories of movements, and motion processes) or for invisible movements (e.g., on molecular or atomic levels) (Casey, 1996; Flemming, Hart, & Savage, 2000; Park, 1994; Williams & Abraham, 1995). An important feature of animations, as well as static images, is their fidelity level that characterizes the degree of realism or resemblance to the real world. High fidelity levels with many non-essential details that may distract learner attention may not always be instructionally effective, especially for novice learners (Park, 1994; Romiszowsky, 1993). Schnotz and Grzondziel (1996) found that schematic low-fidelity illustrations were retained better than analogical high-fidelity illustrations.

Even though animated visualizations could be appropriate for presenting dynamic information, no convincing evidence has been obtained indicating higher learning effects of animations as compared to static diagrams (Park & Hopkins, 1993). Tversky, Morrison, & Bétrancourt (2002) noted that in order to effectively convey information about complex systems, graphics should conform to the congruence principle. According to this principle, the content and format of the graphics should correspond to the content and format of the concepts to be conveyed, for example, animated graphics should be effective to represent change over time. However, when conveying complex systems that change over time, animations as sequences of discrete steps could be too complex or fast to be accurately perceived. Lowe (2003) suggested that perceptually salient features of animations may distract learner attention from relevant elements of information. Tversky et al. (2002) associated possible ineffectiveness of animations with their transitory nature that makes it difficult for learners to perceive the depicted changes simultaneously.

Hegarty, Kriz, & Cate (2003) examined the effects of static diagrams and computer animations on learner mental models of a mechanical system. Students viewed static diagrams, predicted motion from static diagrams, viewed computer animations, and viewed static and animated diagrams accompanied by verbal commentaries. No evidence was obtained that animated diagrams led to superior understanding compared to static diagrams. Koroghlanian and Klein (2004) demonstrated that participants who studied animation as opposed to static diagrams spent more time on studying the information, with no corresponding gains in learning. Lowe (1999, 2003) noted that learners need to extract relevant information form animation and integrate it into their available knowledge structures in order to build comprehensive mental models. Perceptually salient but irrelevant features of animations may distract learners from more subtle but relevant features.

A number of studies investigated interactions between levels of learner expertise and the effectiveness of dynamic visualizations (simulations and animations) compared to traditional static visual representations. Kalyuga (2008a) studied the interaction between levels of learner expertise and effectiveness of animated and static procedural examples in the task domain of transforming graphs of linear and quadratic equations in mathematics. The results demonstrated that less knowledgeable learners performed significantly better after studying static examples. Learners with higher levels of prior knowledge showed better results after studying animated instructions. As levels of learner expertise increased, the performance of the animated instruction group improved more than performance of the static group.

Schnotz and Rasch (2005) compared effects of animated and static pictures about time phenomena related to the Earth rotation on learners with different levels of learning prerequisites (a combination of prior knowledge scores and intelligence measures). Two different animated pictures were investigated: visual simulation of changes over time when (simulation picture) and an interactive picture that allowed students to manipulate the display by entering specific day and time for specific cities (manipulation picture).

For simulation pictures, the results indicated that high learning prerequisite learners spent more time on studying animated than static representations, with a reversed pattern for low learning prerequisite students. Students with low learning prerequisites performed significantly better after learning with static pictures than with animated pictures, while high learning prerequisite students performed equally in both conditions. These students were able to perform mental simulations without redundant for them external support. For manipulation pictures, students with high learning prerequisites performed significantly better after learning from manipulation pictures than from simulation pictures, while lower learning prerequisite students performed better after learning from simulation pictures than from manipulation pictures.

According to cognitive load theory, continuous animations could be too cognitively demanding for novice learners because of high levels of transitivity, and these learners would benefit more from studying a set of static diagrams. Knowledge structures of more experienced learners may help them to handle the transitivity of animations, but processing details in static graphics may require redundant activities for these learners. Similarly, manipulation pictures could impose extraneous load on novice learners but be optimal for more experienced learners

As the level of learner expertise in a domain increases, more domain-specific knowledge structures are acquired in long-term memory. These structures allow chunking of many elements of information into familiar units that are treated as single elements in working memory. As a result, more working memory resources become available for constructing and running dynamic mental representations. Static graphics may be less beneficial for more experienced learners than for novices because their available dynamic knowledge structures would need to be integrated and reconciled with redundant for them details displayed in graphics. Additional cognitive resources may be required for such processes, increasing working memory demands and reducing relative learning effects.

Expertise Reversal in Hypertext and Hypermedia Learning Environments

The level of learner prior domain-specific knowledge is an important factor influencing the effectiveness of hypermedia learning environments. It could be difficult for novice learners to maintain goals in working memory while searching for relevant information. The search processes may consume resources that would become unavailable for constructing relevant knowledge structures. Variety of possible navigational choices may increase extraneous cognitive load. Learners with lower levels of prior knowledge may need adequate forms of instructional support to be provided to them. Unsupported hypermedia environments could be suitable for more experienced learners. The available knowledge base would guide these learners in their exploration of the environment (Gerjets & Scheiter, 2007).

Simons and Klein (2007) studied how embedded scaffolds and student achievement levels influence learning in a problem-based hypermedia learning environment (an inquiry project on designing and planning a balloon trip around the world for seventh grade students). Three conditions were investigated: (1) exploratory problem-based environment without scaffolding; (2) an optional scaffolding condition; and (3) a compulsory scaffolding condition that required all students to study all included scaffolds. Results indicated that no scaffolding condition was significantly worse than both scaffolding conditions; high prior achieving students benefited more from the exploratory problem-based environment than low prior achieving student. Analyses of student project notebooks revealed that students in the compulsory scaffolding condition made more directly relevant entries, thus demonstrating qualitative differences in favor of compulsory scaffolds.

Tergan (1997) noted that there were certain rationales favouring multiple representations in hypertext/hypermedia systems, such as corresponding to principles of encoding specificity and encoding variability, supporting the construction of context-indexed mental representations, providing conditions for multimodal cognitive processing and for individualizing instruction. However, the number of representations or perspectives used in an instructional environment does not automatically improve learning outcomes. Sufficient instructional scaffolding is required, especially for novice learners who may face increased cognitive load while integrating several representations and establishing interrelationships between them. The level of learner prior domain-specific knowledge is a major factor that may influence learning in non-linear hypertext or hypermedia learning environments (Dillon & Gabbard, 1998; Shapiro & Niederhauser, 2004). Studies of the role of prior knowledge in learning from hypertext have generally indicated that low prior knowledge learners benefit more from well structured linear texts. Even though higher prior knowledge learners are usually expected to benefit from non-linear environments and network structures (e.g., Spiro,

Feltovich, Jacobson & Coulson, 1991; Jacobson, 1994), the sufficient positive evidence in support of this suggestion still needs to be accumulated.

The general cognitive load assumption about the role of prior knowledge in learning from hypertext is that only relatively knowledgeable learners could make sense of unordered segments of text without overloading working memory. These learners may handle associated split attention situations by using their existing knowledge without cognitive overload (DeStefano & LeFevre, 2007). For example, a limited-access version of hypertext (with restricted navigation facilities) was more beneficial for low prior knowledge learners than a full access version (Shin, Schallert, & Savenye, 1994). There was no difference between these two hypertext versions for students with higher levels of prior knowledge in the topic

On the other hand, Calisir and Gurel (2003) demonstrated that while for non-knowledgeable learners, two hypertext structures (hierarchical and mixed) were significantly better than linear text, results reversed for knowledgeable learners. Domain knowledge may have helped these learners to understand the structure of the text. There were no significant differences between knowledgeable and non-knowledgeable learners in the hierarchical and the mixed conditions. Well structured hierarchical and mixed forms of hypertext enhanced comprehension for less-knowledgeable learners by compensating for the lack of internal conceptual structure of the domain.

Potelle and Rouet (2003) compared three levels of hypertext organization: a hierarchical map with the most explicit structure; a semantic network map; and an alphabetical list of topics. Low-knowledge readers learned global social psychology concepts more from the hierarchical map than from other two structures, while there were no significant differences for high-knowledge readers. Hierarchical representations better facilitated the construction of explicit representations and the integration of individual topics for low-knowledge learners. Shapiro (1999) found that the explicit external structuring by interactive overviews in hypermedia-based learning environments produced significant benefits for novices, however made little difference for learners with higher levels of prior familiarity with the domain for whom the explicit external structures were not necessary.

Thus, while for more experienced learners, both hypertext/hypermedia and linear text environments may work well, for novices the answer may depend on the specific nature of the learning material. While the linear text is a preferred default option, well structured hypertext may represent the content better than traditional linear text and could become beneficial for novice learners. Characteristics of hypertext/hypermedia learning materials that make them suitable for representing specific task domains in a cognitively optimal way still need to be investigated.

Adaptive Multimedia Learning Based on the Expertise Reversal Effect

A major instructional implication of the expertise reversal effect is the need to tailor dynamically instructional techniques, procedures, and levels of instructional guidance to current levels of learner expertise as they gradually change during learning. Changes in the learner's knowledge base need to be dynamically tracked and specific instructional techniques and procedures tailored accordingly. As levels of learner expertise increase, relatively less-guided exploratory, problem-solving, or game-base environments could effectively assist in learning advanced knowledge and skills in specific task domains.

The research on cognitively-optimized adaptive strategies within a cognitive load framework is still very limited. In the absence of comprehensive research- and evidence-based recommendations, most of existing adaptive environments are based on monitoring learner external characteristics (navigational patterns, learning styles, preferences etc.) rather than deep cognitive characteristics, such as available knowledge base and levels of learner expertise.

The ability to diagnose levels of learner expertise rapidly and in real time is an important prerequisite to building dynamically-tailored learning environments. Such information could be obtained by observing how learners approach briefly presented tasks. This general idea was implemented in the first-step diagnostic method: learners were presented with selected tasks for a limited time and asked to rapidly indicate their first step towards solution of each task. Well learned solution procedures would allow more experienced learners to rapidly generate advance steps of the solution and skip some intermediate steps. Different first steps could be indicators of different levels of expertise. This technique was validated in a series of studies that indicated significant correlations (up to .92) between performance on the rapid tasks and traditional measures of knowledge (Kalyuga, 2006d; Kalyuga & Sweller, 2004), with test times reductions by factors of up to 4.9.

In an alternative rapid testing method, learners were presented with a series of potentially possible steps at various stages of the solution procedure, and asked to rapidly verify the correctness of these steps. This rapid verification method is also potentially usable for relatively poorly defined task areas when solution steps could not be specified precisely (e.g., when the solution procedure requires drawing graphical representations or when there are several possible solution paths). The method was validated using sentence comprehension, kinematics, and mathematics tasks, and indicated a significant correlation (up to .75) between performance on rapid tasks and traditional measures of expertise, with test time reduced by factor up to 3.7 (Kalyuga, 2006c; 2008b).

The suggested diagnostic methods were used in adaptive online tutorials in the domains of linear algebra equations (Kalyuga & Sweller, 2004; Kalyuga & Sweller, 2005) and vector addition motion problems in kinematics (Kalyuga, 2006a). According to the dynamic tailoring approach, the tutorials provided levels of instructional guidance that were optimal for learners with different levels of expertise as measured by online methods. In learner-adapted groups, at the beginning of training sessions, each learner was provided with an appropriate level of instructional guidance according to the outcome of the initial rapid pretest. Depending on the outcomes of the ongoing rapid tests during the session, the learner was allowed to proceed to the next learning stage or was required to repeat the same stage and then take the rapid test again. At each subsequent stage, a lower level of guidance was provided to learners (e.g., worked-out components of solution procedures were gradually omitted and progressively replaced with problem solving steps), and a higher level of the rapid diagnostic tasks was used at the end of the stage.

In Kalyuga & Sweller (2005) and Kalyuga (2006a), the rapid measures of expertise were combined with measures of cognitive load (using subjective ratings of task difficulty). Since the expertise is associated not only with relatively higher-level but also lower-effort performance, combining both measures was expected to produce a better indicator of learner expertise in a domain. Results indicated that both adaptive conditions outperformed the nonadapted condition group, however, there were no significant differences between the adaptation procedures. These results supported previous results obtained by Camp, Paas, Rikers, & van Merriënboer (2001), Salden, Paas, Broers, & van Merriënboer (2004), and Salden, Paas, & van Merriënboer (2006) in air traffic control training.

The above approaches to dynamic tailoring of instruction to learner characteristics assume a system-controlled format: a computer program or instructor dynamically select an instructional method that is most appropriate for the current level of learner expertise. An alternative format could be realizes as a learner-controlled approach to the individualization of instruction. However, despite expected motivational advantages (e.g., positive learner attitudes and a sense of control) and higher learning outcomes of learner control, research findings have been inconclusive and more often negative rather than positive in relation to learning outcomes (Chung & Reigeluth, 1992; Niemec, Sikorski, & Walberg, 1996). They indicate that learners may require assistance in effective use of provided control facilities.

According to cognitive load theory, the level of learner expertise is a defining factor: students could have control over the content and instructional sequences when they have sufficient knowledge of the task domain; otherwise they may require appropriate assistance. One form of such assistance is providing guidance and advisement to learners for making their own decisions (Tennyson & Rothen, 1979). Using this information, learners can make effective decisions themselves. An advanced contemporary form of this approach is an adaptive guidance strategy: providing learners with individualized prescriptive information on the current level of their knowledge, recommended learning materials and tasks based on past performance, how to sequence learning tasks, and how to allocate cognitive resources (Kozlowski, Toney, Mullins, Weissbein, Brown, & Bell, 2001). Based on an adaptive guidance approach, Bell and Kozlowski (2002) demonstrated that providing students with adaptive guidance in addition to learner control in a complex learning environment improved the acquisition of basic knowledge and skills for novice learners and strategic knowledge and transfer capabilities for more advanced students.

Shared-control models as alternatives to system-controlled task selection models were suggested within a cognitive load framework (Van Merriënboer, Sluijsmans, Corbalan, Kalyuga, Paas, & Tattersall, 2006). For example, Corbalan, Kester, and van Merriënboer (2006) investigated a shared control model that first selected a subset of tasks based on learner performance scores and cognitive load ratings (a system-controlled component), and then presented this subset to the learner who made the final decision (a learner-controlled component). Results indicated that learners in the shared control condition demonstrated higher posttest scores with lower cognitive load than learners in the system-controlled condition.

CONCLUSION

The expertise reversal effect has been observed in many studies within and outside of a cognitive load framework in multimedia and hypermedia learning environments. The need for more knowledgeable learners to integrate and cross-reference redundant instructional guidance with available knowledge structures that relate to the same situations may consume additional cognitive resources. A minimal instructional guidance would allow these learners to take advantage of their knowledge base in the most efficient way. Instructional guidance should be provided at the appropriate time, while unnecessary support removed as a learner progresses to more advanced levels of proficiency in a specific domain. Adaptive learning environments that dynamically tailor levels of instructional support to changing individual levels of learner expertise in a domain have the best potential for optimizing cognitive load.

Existing adaptive multimedia environments are mostly based on external characteristics of learner behavior rather than on real cognitive characteristics. Learner levels of prior knowledge or expertise should be made the primary factors in adaptive instructional systems, complemented with relevant secondary factors (e.g., navigational patterns, learning styles, and preferences). Individualized adaptive instruction should be based on detailed diagnostic assessment of learner knowledge structures that could appropriately direct instructional interventions. Different variable levels of learner control could be implemented in these environments as means of enhancing their adaptive capabilities. Adaptive guidance approaches could be used to monitor and assess learner progress and provide learners with diagnostic information and individual recommendations on future learning activities. As learners acquire basic lower-level knowledge and skills, adaptive guidance should tailor subsequent learning tasks and activities to focus attention on more advanced knowledge and skills. A continuously available instructional support is important even when adaptive advice is provided to learners.

The quality of adaptive environments depends on the accuracy of information about levels of learner knowledge and skills. Using traditional tests and tracing user interactions with the system could be imprecise and incomplete. Applying modern artificial intelligence approaches and using fine-grained production rule-based learner models in intelligent tutoring systems (e.g., Anderson, Corbett, Fincham, Hoffman, & Pelletier, 1992) allowed a significant increase in the precision of adaptive methodologies. However, implementations of such approaches require complex computational modeling and, therefore, have been limited to several well defined domains (e.g., programming and mathematics). On the other hand, most adaptive hypermedia and web-based environments are based on few discrete coarse-grained levels of user expertise (e.g., high, intermediate, low levels; De Bra & Calvi, 1998).

A rapid diagnosis-based approach may offer appropriate tools that combine high levels of diagnostic precision with simplicity of implementation in learner-tailored instructional procedures. Rapid diagnostic instruments for measuring levels of learner expertise in poorly defined task areas will be required for the development of adaptive learning environments in such domains. The expertise reversal effect may provide a valuable guidance to instructional designers, particularly for the design of learner-adapted multimedia instructional systems.

REFERENCES

Anderson, J. R., Corbett, A. T., Fincham, J. M., Hoffman, D. & Pelletier, R. (1992). General principles for an intelligent tutoring architecture. In V. Shute and W. Regian (Eds.), *Cognitive approaches to automated instruction* (81-106). Hillsdale, NJ: Erlbaum.

Baddeley, A. D. (1986). *Working Memory*. New York: Oxford University Press.

Bell, B. S. & Kozlowski, S. W. J. (2002). Adaptive guidance: Enhancing self-regulation, knowledge, and performance in technology-based training. *Personnel Psychology., 55,* 267-306.

Calisir, F. & Gurel, Z. (2003). Influence of text structure and prior knowledge of the learner on reading comprehension, browsing and perceived control. *Computers in Human Behavior., 19,* 135-145.

Camp, G., Paas, F., Rikers, R. & van Merriënboer, J. J. G. (2001). Dynamic problem selection in air traffic control training: A comparison between performance, mental effort, and mental efficiency. *Computers in Human Behavior., 17,* 575-595.

Casey, C. (1996). Incorporating cognitive apprenticeship in multi-media. *Educational Technology Research and Development., 44,* 74-84.

Chandler, P. & Sweller, J. (1991). Cognitive load theory and the format of instruction. *Cognition and Instruction., 8,* 293-332.

Chung, J. & Reigeluth, C. M. (1992). Instructional prescriptions for learner control. *Educational Technology., 32,* 14-20.

Corbalan, G., Kester, L. & van Merriënboer, J. J. G. (2006). Towards a personalized task selection model with shared instructional control. *Instructional Science, 34.*, 399-422.

Cowan, N. (2001). The magical number 4 in short-term memory: A reconsideration of mental storage capacity. *Behavioral and Brain Sciences., 24,* 87-114.

De Bra, P. & Calvi, L. (1998). AHA! An open Adaptive Hypermedia Architecture. *The New Review of Hypermedia and Multimedia., 4,* 115-139.

DeStefano, D. & LeFevre, J-A. (2007). Cognitive load in hypertext reading: A review. *Computers in Human Behavior., 23,* 1616-1641.

Dillon, A. & Gabbard, R. (1998). Hypermedia as an educational technology: A review of the quantitative research literature on learner comprehension, control, and style. *Review of Educational Research., 68,* 322-349.

Flemming, S. A., Hart, G. R. & Savage, P. B. (2000). Molecular orbital animations for organic chemistry. *Journal of Chemical Education., 77,* 790-793.

Gerjets, P. & Scheiter, K. (2007). Learner control in hypermedia environments. *Educational Psychology Review., 19,* 285-307.

Hegarty, M., Kriz, S. & Cate, C. (2003). The roles of mental animations and external animations in understanding mechanical systems. *Cognition and Instruction., 21,* 325-360.

Jacobson, M. J. (1994). Issues in hypertext and hypermedia research: Toward a framework for linking theory-to-design. *Journal of Educational Multimedia and Hypermedia., 3(2),* 141-154.

Kalyuga, S. (2005). Prior knowledge principle. In R. Mayer (Ed.), *Cambridge Handbook of Multimedia Learning.*, (325-337). New York: Cambridge University Press.

Kalyuga, S. (2006a). Assessment of Learners' Organized Knowledge Structures in Adaptive Learning Environments. *Applied Cognitive Psychology., 20,* 333-342.

Kalyuga, S. (2006b). *Instructing and testing advanced learners: A cognitive load approach..* New York: Nova Science.

Kalyuga, S. (2006c). Rapid assessment of learners' proficiency: A cognitive load approach. *Educational Psychology., 26,* 613-627.

Kalyuga, S. (2006d). Rapid cognitive assessment of learners' knowledge structures. *Learning and Instruction., 16,* 1-11.

Kalyuga, S. (2007). Expertise reversal effect and its implications for learner-tailored instruction. *Educational Psychology Review., 19*, 509-539.

Kalyuga, S. (2008a). Relative effectiveness of animated and static diagrams: An effect of learner prior knowledge. *Computers in Human Behavior., 24,* 852-861.

Kalyuga, S. (2008b). When less is more in cognitive diagnosis: A rapid assessment method for adaptive learning environments. *Journal of Educational Psychology, 100,* 603-612.

Kalyuga (2009). *Managing cognitive load in adaptive multimedia learning.,* Hershey – New York: Information Science Reference.

Kalyuga, S., Chandler, P. & Sweller, J. (1998). Levels of expertise and instructional design. *Human Factors., 40*, 1-17.

Kalyuga, S., Chandler, P. & Sweller, J. (2000). Incorporating learner experience into the design of multimedia instruction. *Journal of Educational Psychology., 92*, 126-136

Kalyuga, S. & Sweller, J. (2004). Measuring knowledge to optimize cognitive load factors during instruction. *Journal of Educational Psychology., 96*, 558-568.

Kalyuga, S. & Sweller, J. (2005). Rapid dynamic assessment of expertise to improve the efficiency of adaptive e-learning. *Educational Technology Research and Development., 53*, 83-93.

Koroghlanian, C. & Klein, J. D. (2004). The effect of audio and animation in multimedia instruction. *Journal of Educational Multimedia and Hypermedia., 13*, 23-46.

Kotovsky, K., Hayes, J. R. & Simon, H. A. (1985). Why are some problem hard? Evidence from Tower of Hanoi. *Cognitive Psychology., 17*, 248-294.

Kozlowski, S. W. J., Toney, R. J., Mullins, M. E., Weissbein, D. A., Brown, K, G. & Bell, B. S. (2001). Developing adaptability: A theory for the design of integrated-embedded training systems. In E. Salas (Ed.), *Advances in human performance and cognitive engineering research* (Vol. 1) (59-123). Amsterdam: Elsevier Science.

Lambiotte, J. G. & Dansereau, D. F. (1992). Effects of knowledge maps and prior knowledge on recall of science lecture content. *Journal of Experimental Education., 60*, 189-201.

Lee, H., Plass, J. L. & Homer, B. D. (2006). Optimizing cognitive load for learning from computer-based science simulations. *Journal of Educational Psychology., 98*, 902-913.

Lowe, R. K. (1999). Extracting information from an animation during complex visual learning. *European Journal of Psychology of Education., 14*, 225-244.

Lowe, R. K. (2003). Animation and learning: Selective processing of information in dynamic graphics. *Learning and Instruction., 13*, 157-176.

Mayer, R. E. (1997). Multimedia learning: Are we asking the right questions? *Educational Psychologist., 32*, 1-19.

Mayer, R. & Anderson, R. (1991). Animations need narrations: An experimental test of a dual-coding hypothesis. *Journal of Educational Psychology., 83*, 484-490.

Mayer, R. & Anderson, R. (1992). The instructive animation: Helping students build connections between words and pictures in multimedia learning. *Journal of Educational Psychology., 84*, 444-452.

Mayer, R. & Gallini, J. (1990). When is an illustration worth ten thousand words*? Journal of Educational Psychology., 82*, 715-726.

Mayer, R., Heiser, J. & Lonn, S. (2001). Cognitive constraints on multimedia learning: When presenting more material results in less understanding. *Journal of Educational Psychology., 93*, 187-198.

Mayer, R. E. & Sims, V. K. (1994). For whom is a picture worth a thousand words? Extensions of a dual-coding theory of multimedia learning. *Journal of Educational Psychology., 86*, 389-401.

Mayer, R. E., Steinhoff, K., Bower, G. & Mars, R. (1995). A generative theory of textbook design: Using annotated illustrations to foster meaningful learning of science text. *Educational Technology Research and Development., 43*, 31-43.

Miller, G. A. (1956). The magical number seven, plus or minus two: Some limits on our capacity for processing information. *Psychological Review., 63*, 81-97.

Mousavi, S. Y., Low, R. & Sweller, J. (1995). Reducing cognitive load by mixing auditory and visual presentation modes. *Journal of Educational Psychology., 87*, 319-334.

Niemec, P., Sikorski, C. & Walberg, H. (1996). Learner-control effects: A review of reviews and a meta-analysis. *Journal of Educational Computing Research., 15*, 157-174.

Park, O. (1994). Dynamic visual displays in media-based instruction. *Educational Technology.*, 21-25.

Park, O-C. & Hopkins, R. (1993). Instructional conditions for using dynamic visual displays. *Instructional Science., 21*, 427–449.

Pollock, E., Chandler, P. & Sweller, J. (2002). Assimilating complex information. *Learning and Instruction., 12*, 61-86.

Potelle, H. & Rouet, J. F. (2003). Effects of content representation and readers' prior knowledge on the comprehension of hypertext. *International Journal of Human-Computer Studies., 58*, 327-345.

Rieber, L. P. (1990). Animation in computer-based instruction. *Educational Technology, Research and Development., 38*, 77-86.

Romiszowsky, A. J. (1993). Psychomotor principles. In M. Fleming, & W. H. Levie (Eds.), *Instructional message design: principles from the behavioral and cognitive sciences* (127-189). Englewood Cliffs, NJ: Educational Technology Publications.

Salden, R. J. C. M., Paas, F., Broers, N. J. & van Merriënboer, J. J. G. (2004). Mental effort and performance as determinants for the dynamic selection of learning tasks in air traffic control training. *Instructional Science., 32*, 153-172.

Salden, R. J. C. M., Paas, F. & van Merriënboer, J. J. G. (2006). Personalized adaptive task selection in air traffic control: Effects on training efficiency and transfer. *Learning and Instruction., 16*, 350-362

Schnotz, W. & Grzondziel, H. (1996). *Knowledge acquisition with static and animated pictures in computer-based learning*. Paper presented at the annual meeting of the American Educational Research Association, New York City. (ERIC No. ED 401 878)

Schnotz, W. & Rasch, T. (2005). Enabling, facilitating, and inhibiting effects of animations in multimedia learning: Why reduction of cognitive load can have negative results on learning. *Educational Technology Research and Development., 53*, 47-58.

Shapiro, A. M. (1999). The relationship between prior knowledge and interactive overviews during hypermedia-aided learning. *Journal of Educational Computing Research., 20*, 143-167.

Shapiro, A. & Niederhauser, D. (2004). Learning from Hypertext: Research issues and findings. In D. Jonassen (Ed.), Handbook of Research for Educational Communications and Technology (605-620). Hillsdale, NJ: Lawrence Erlbaum Associates.

Shiffrin, R. & Schneider, W. (1977). Controlled and automatic human information processing: II. Perceptual learning, automatic attending, and a general theory. *Psychological Review., 84*, 127-190.

Shin, E. C., Schallert, D. L. & Savenye, W. C. (1994). Effects of learner control, advisement, and prior knowledge on young students' learning in a hypertext environment. *Educational Technology Research and Development., 42*, 33-46.

Simons, K. D. & Klein, J. D. (2007). The impact of scaffolding and student achievement levels in a problem-based learning environment. *Instructional Science., 35*, 41-72.

Spiro, R. J., Feltovich, P. J., Jacobson, M. J. & Coulson, R. L. (1991). Knowledge representation, content specification, and the development of skill in situation-specific knowledge assembly: Some constructvist issues as they relate to cognitive flexibility theory and hypertext. *Educational Technology., 31* (9), 22-25.

Sweller, J., Chandler, P., Tierney, P. & Cooper, M. (1990). Cognitive load and selective attention as factors in the structuring of technical material. *Journal of Experimental Psychology: General., 119*, 176-192.

Tergan, S. (1997). Multiple views, contexts, and symbol systems in learning with hypertext/hypermedia: a critical review of research. *Educational Technology.*, July-August, 5-18.

Tennyson, R. D. & Rothen, W. (1979). Management of computer-based instruction: Design of an adaptive control strategy. *Journal of Computer-Based Instruction., 5*, 126-134.

Tindall-Ford, S., Chandler, P. & Sweller, J. (1997). When two sensory modes are better than one. *Journal of Experimental Psychology: Applied., 3(4)*, 257-287.

Tversky, B., Morrison, J. B. & Betrancourt, M. (2002). Animation: Does it facilitate learning? *International Journal of Human-Computer Studies., 57*, 247-262.

Van Merriënboer, J. J. G., Sluijsmans, D., Corbalan, G., Kalyuga, S., Paas, F. & Tattersall, C. (2006). Performance assessment and learning task selection in environments for complex learning. In D. Clark, and J. Elen (Eds.), *Advances in Learning and Instruction.,* (201-220). Amsterdam: Elsevier Science.

Williams, V. M. & Abraham, M. R. (1995). The effects of computer animation on the particulate mental models of college chemistry students. *The Journal of Research in Science Teaching., 32*, 521-534.

In: Computer-Assisted Teaching: New Developments
ISBN: 978-1-60876-855-4
Editors: B. A. Morris, G. M. Ferguson, pp. 133-149

Chapter 6

COMPARISON OF MULTIMEDIA COMPUTER-ASSISTED INSTRUCTION, TRADITIONAL INSTRUCTION AND COMBINED INSTRUCTION ON KNOWLEDGE ACQUISITION AND RETENTION OF SETTING SKILL IN VOLLEYBALL

Nicholas Vernadakis, Panagiotis Antoniou, Eleni Zetou, Maria Giannousi and Eftimis Kioumourtzoglou

Democritus University of Thrace, Department of Physical Education and Sport Science, Komotini, Greece

ABSTRACT

The purpose of this study was to examine the effect of multimedia computer - assisted instruction (MCAI), traditional instruction (TI), and combined instruction (CI) methods on learning the skill of setting in Volleyball. Additionally, a comparison of the students' attitudes towards the MCAI and TI methods was made. Forty-eight middle school students of seventh and eighth grade were randomly assigned into three teaching method groups: TI, MCAI and CI. Each group received ten 45-min periods of instruction divided into 3 sections: a) 5-min introduction, b) 30-min instructional time and c) 10-min questions and review. Students took pre-, post-, and retention written test covering rules and strategy of the games. Participants in the CI group also completed a post-test attitude survey towards the MCAI and TI methods. Two-way analysis of variances (ANOVA), with repeated measures on the last factor, were conducted to determine effect of method groups (MCAI, TI, CI) and measures (pre-test, post-test, re-test) on knowledge test. Paired samples t-test analyses were conducted to measure students' attitude towards the MCAI and TI methods. Post-test results indicated no significant differences between the groups concerning the written test. Nevertheless, the attitude test scores of the CI group were more favourable to MCAI method than the TI method. Retention test results showed that groups retained the knowledge acquisition. However, the combine method of instruction tended to be the most effective on cognitive learning.

Keywords: Instructional technology, multimedia software, traditional instruction, cognitive learning, attitude, volleyball.

INTRODUCTION

The importance of visual information in the learning process is well known. Consider that, on average, people learn 1% through what they taste, 1.5 % through what they touch, 3.5 % through what they smell, 11 % through what they hear, but 83 % through what they see. Research also finds that individuals retain 10 % of what they write, 20 % of what they hear, 30 % of what they see, 50 % of what they see and hear, and 70% of what they discuss (Simatos, 2000). Based upon these data, the future of learning points in the direction of interactive, multimedia computer assisted instruction (MCAI), and the ideal delivery medium for the masses is the next-generation broadband Internet.

So what then is "interactive multimedia?" Although there are many definitions, Akhtar (2003) described it as using many different media (print, audio, video, etc.) to present more comprehensive information than any medium alone can, accommodating children with different learning styles, and employing interactivity to stimulate children to become active, motivated learners.

The use of multimedia software is becoming more popular with the improvement in computer technology. But should it be used because it is popular or because it meets the instructional objectives? There is debate going on about the effectiveness of multimedia computer-assisted instruction (MCAI) versus the traditional instruction (TI) approach or teacher-led approach. The literature reports a considerable amount of research on the use of multimedia by educators and on their effectiveness in the learning process. Blok, Oostdam, Otter, & Overmaat (2002), for example, examined the effectiveness of computer-assisted instruction (CAI) programs in supporting beginning readers. Their review included 42 studies from 1990 onward, and they found the corrected overall effect size estimate was .19. Their findings were similar to earlier meta-analyses by Kulik and Kulik (1991) and Bangert-Drowns (1993), which also examined the effects of CAI and found it to have positive but small effects. Wiemeyer, (2003) reviewed nine meta-analyses of earlier and different multimedia issues and suggested that multimedia learning can be more effective and efficient than traditional learning. But this effect depends on many factors like the features of the learners, the teachers, the learning stuff, the type of learning, the features of the study, etc. Further, a meta-analysis of 167 studies (Bernard, Abrami, Lou, Borokhovski, Wade, Wozney, Wallet, Fiset, & Binru Huang, 2004) concluded that a very weak learning advantage for multimedia in empirical studies is based on uncontrolled instructional methods.

One area in which there have not been many meta-analyses and systematic reviews of the research is how teaching and learning with multimedia technology impacts student outcomes in physical education. This area is important because some studies have found that multimedia technology can change teachers' pedagogic practices from a teacher-centered or teacher-directed model to a more student-centered classroom where students work cooperatively, have opportunities to make choices, and play an active role in their learning. Antoniou, Derri, Kioumourtzoglou, & Mouroutsos (2003), for example, examined the effect of multimedia computer assisted instruction (MCAI), traditional instruction (TI), and combined instruction (CI) on learning rule violations in basketball by university physical education students. Written test results indicated that students in all groups improved their knowledge of rule violations but only those in the TI and CI groups retained this knowledge. Also, the researchers found that TI group showed significantly greater retention than the MCAI group both in the written test and in total performance.

Wiksten, Spanjer, & LaMaster (2002), investigated the effectiveness of MCAI in an introductory athletic training laboratory class as a supplement to TI approach. The results showed no significant differences between groups on either the written or practical examination. Student attitudes toward the CD-ROM software were favorable, and the qualitative data suggest that students would use this type of educational resource. In another study that examined changes in physical activity and nutritional patterns of high school physical education students as a result of multimedia technology, Everhart, Harshaw, Everhart, Kernodle, & Stubblefield (2002) found that the year-long multimedia intervention did not affect physical activity or nutritional patterns of students significantly.

Vernadakis, Zetou, Antoniou, & Kioumourtzoglou (2002) reported that MCAI is a functional method in teaching the skill of volleyball setting to children aged 12 – 14 year old and is as effective as traditional teaching. The results of this study showed that there were no significant differences between the TI and MCAI groups with regards to the knowledge and skill test. In another study Wilkinson, Hillier, Padfield, & Harrison (1999) found that junior high school girls in both classes (TI and MCAI) significantly improved in their knowledge of volleyball rules and in performance of most skills (pass, set, and underhand serve) during the 16-day unit. The MCAI class had significantly better scores on the forearm pass, more successful passes/serve, set/serve, total contacts/serve and more sets than the TI class. However, no significant differences were found between the two classes in knowledge of rules. Survey results revealed that the majority of the treatment class felt that using the multimedia software helped improve their motor and cognitive skills.

More recently, Goran and Reynolds (2005) examined the efficacy of a computer-based interactive multimedia curriculum for promoting physical activity in fourth grade children. The researchers concluded that the interactive multimedia curriculum favored an improvement in obesity indices in girls and was associated with subtle changes in physical activity in girls and general improvement in psychosocial outcomes related to physical activity. In another study that examined the effective use of an Interactive Instructional Multimedia Software in the teaching process of alpine skiing, Antoniou, Moulelis, Siskos, & Tsamourtzis (2006) found that the multimedia assisted instruction method of teaching alpine skiing in beginners is more effective than the traditional method.

Finally, Si-min Li and Jin-hai Sun (2008) in an experimental research on multimedia teaching for sports aerobics found that: (1) the multimedia teaching for sports aerobics, which takes the students' study as the center, pays great attention to the learning environment design is helpful in making the student to establish the correct technical movement concept, and then raise the utilization rate of effective time in class, and increase the teaching information capacity the grades, (2) the sports aerobics received in the experiment group are better than those of the students in the opposite one, and (3) the multimedia teaching has its unique superiority in theoretical knowledge and the technical skill instruction aspects of sports aerobics compared to conventional teaching methods.

Taking these factors into consideration, it would be interesting to see the effectiveness of MCAI in a learning environment, more specifically in a physical education learning environment and then to point out the lacuna that needs to be filled. But first, it is important to find out how students feel about it as it is they who will be instructed through the whole gamut of the learning process.

Therefore, the purpose of this study was to compare three different instructional methods (MCAI, TI & CI) by means of the knowledge test scores, obtained from three groups of middle school students. Additionally, a comparison of the students' attitudes towards the MCAI and TI methods was made. The tests assessed the learning of the setting skill in volleyball. More specifically, the study was conducted to explore the following four research questions:

1. Should one or more items on knowledge test be deleted or revised to obtain a better measure of setting skill in volleyball?
2. Do students, on average, report differently on knowledge test using the TI, the MCAI and the CI teaching approaches?
3. Do students, on average, report differently on knowledge test for the pre-test, post-test and re-test measurements?
4. Do the differences in means for knowledge test between the TI, MCAI and CI teaching method groups vary between the pre-test, post-test and re-test measurements?
5. Are students more favourable on the average about TI or MCAI teaching approaches?

METHOD

Participants

To obtain permission for conducting the field experiment, the researchers contacted local middle schools in a northern city of Greece. All school principals expressed their willingness to participate. The private school Dellasal of Thessaloniki, having an indoor gymnasium and essential network equipment, was chosen for the experiment. Forty-eight (n=48) middle school students (25 girls and 23 boys) of seventh and eighth grade, 12-14 years of age (M=13, S.D. =1.01), selected for this study by random sampling method, enrolled in the volleyball course. Participants were randomly assigned to one of the three different teaching methods: TI, CI and MCAI creating three independent groups of 16 students. All participants had no formal training on learning the skills of setting in volleyball. Prior to group assignments, participants were orientated to the purpose of the study and participant requirements. Following the orientation, informed consent was obtained from each participant. The students should have returned the informed consent form signed by his/her parent or guardian in order to participate in the research.

Apparatus

Hardware. Ten 3.2 GHz Intel Pentium 4 class computers with an 800 MHz front side bus and Intel's new Hyper-Threading technology for increased performance during demanding operations such as complex multitasking running Windows XP professional SP2 were used. The computers equipped with 1024 MB RAM memory, a high-end nVidia GeForce Fx Go 5600 graphics controller with 128 MB of dedicated video memory, a 17-inch color LCD monitor, DVD-ROM, soundcard, microphone and small headset.

Software. A multimedia program was created and programmed in Asymetrix ToolBook to administer experimental events including 163 screens; 5 screens were introductory, 1 was main menu, 48 were information, 40 were practice, 57 were feedback, and 12 were help. Material for the multimedia application was taken from a volleyball coaching textbook (American Sport Education Program, 2001) and modified for this study. Systematic Instructional Design (SID) concept was used to design the multimedia learning material. The application consisted of four sections: a) history, b) rules, court and player's position, c) skill fundamentals, and d) skill exercises. Two choices menus, one for the termination of the program and one for help, were also included at the bottom of the screen and were always available. The help menu contained a description of the active picture-buttons and suggestions for the program use. The program started with an introductive video of international volleyball federation (FIVB). The main menu with four active picture-buttons which serve as links to the other sections of the program followed.

The first two major sections addressed basic knowledge of the volleyball game pertaining to vocabulary used, history of the game, rules, court dimensions, names of positions, and rotating positions. The skill fundamental and skill exercise sections introduced basic setting skills and exercises for practical work in terms and levels that were appropriated for beginning volleyball players. A step by step instructional format that was accompanied by an exceptional graphic simulation depicting proper form of setting skill at different stage was included in these sections. A discussion of possible errors, what causes these errors, and what may be done to correct these errors was provided in the description of setting skill. This was followed by "animated graphics" that showed the skill in a continuous movement pattern. When the user had seen enough of the setting skill, he could supplement a short quiz (multiple choices, true/false) regarding the technique and concepts that were presented.

Each section included different relevant material like text, sound, pictures, animated vector images, graphics and video. The video presented professional volleyball players performing the setting skills. Each skill was demonstrated several times and was shown from different angles. Close-up shots revealed the details of important points such as hand or foot position. Audio was used to explain each action and give execution cues to help focus the attention of the user. The user navigated through the sections from the menu that appeared on each screen. At the end of the program, a screen with the title of the program, the names of the author and the institution were presented.

Knowledge Test. A knowledge test was developed to determine students' achievement on cognitive learning from the skill of setting in volleyball. A table of specifications was developed to reflect the interrelationship between the identified course content and the levels of learning. Initial drafting of questions for the instrument relied upon survey research handbook by McGee and Farrow (1987). Based on these specifications a 20-item, multiple-choice test was constructed. Each test item had four options in order to reduce the probability of guessing. The test construction was based on the linear model which required that the test scores were obtained by summing the number of correct answers with equal weighting over the items. The questions were written on the basis of the learning objectives outlined in the Ministry of Education's Volleyball Curriculum Guideline.

After the questions were constructed as explained above, a panel of experts in volleyball teaching and coaching was used to evaluate and judge the content validity of the test instrument. This group reviewed the test items and established whether each item measured the target skill. Every time a set of changes was made, the questionnaire was reviewed again by the consultants, until the instrument was deemed adequate.

The revised version of the knowledge test consisted of a 15-item multiple-choice test. Questions included in the knowledge test fell into one of the following categories: a) six skill concepts and b) nine general rules associated with the skill. A pilot study was performed to access item difficulty and clarity of questions. Questions were scored one point for a right answer and no point for a wrong answer.

Attitude Questionnaire. To assess students' attitude, a questionnaire was used consisting of 12 bi-polar items, divided in two sections from the Teaching Method Attitude Questionnaire (TMAQ, Antoniou, Theodorakis, Kioumourtzoglou, 1999). Each section contained six-point Likert scale numeric indicators ranging from "5" (powerful) down to "0" (weak). The first section was referred in TI method and the second in MCAI method with a common point (0). This format allowed participants to select a response from "0" to "5" representing their disagreement or agreement on each section of the particular item respectively (Appendix B). The questionnaire addressed two aspects of attitude: (1) positive (Alpha=.74) and (2) negative (Alpha=.76). The sum scores of positive surnames, minus the sum scores of negative surnames indicated the students' attitude scores in each teaching method (MCAI, TI). The difference of these scores between the two teaching methods (MCAI, TI) represented the students' overall attitude scores. The TMAQ instrument was designed for understanding students' attitude toward MCAI and TI teaching method in a specific situation such as a classroom setting (Antoniou, Theodorakis, Kioumourtzoglou, 1999). This attitude questionnaire was given to CI group as post- test.

Procedure

When the MCAI program was developed, the researchers gave it to an instructional technology specialist, a subject-area expert, and three subject-area teachers for evaluation. Researchers revised and improved the multimedia application according to the feedback received from those experts.

A pilot study was followed to determine the reliability and validity of the knowledge test. Participants consisted of 24 seventh and eighth grade middle school students. This population was chosen to keep the pilot study similar to the main study regarding participant's age. The method of instruction used for the pilot study was TI, which incorporated a direct style of teaching such as lectures, demonstrations, teacher questions and student questions. Participants were given two 45-minute class periods of instruction and review concerning the skill of setting in volleyball. This was done to take into account the fact that participants had not received formal instruction pertaining to this particular skill for almost one year. The knowledge test was administered on the third day on a paper and pencil test consisting of 15 multiple-choice questions. The instruction took place in an indoor gymnasium in order to avoid complications associated with weather conditions.

After the pilot study, a main study was conducted to compare the scores obtained by 48 seventh and eighth grade middle school students for a knowledge test. The experimental design consisted of a pre-test, a post-test and a re-test for the three of the independent groups. The knowledge test was administered on the first day to measure participant's baseline performance on the selected setting skill. Procedures for the knowledge test were the same as the pilot test. There were five fewer questions, reducing the number of questions to ten (Appendix A).

On the third day, ten computers were set up in a blocked-off hallway adjacent to the gymnasium. Each computer had a volleyball skill CD-ROM created by the researchers. Computers were separated as much as possible to create individual workstations. Before the experiment started, the MCAI and CI groups were given a 45-minute introductory session on how to use the multimedia application program prepared for this study. Then the physical education instructor gave a 45-minute lecture to all participants introducing the unit of volleyball. Instruction, practice, and testing for this study were held on six separate and successive weeks. The groups met for 45-minute, 2 times each week in an indoor gymnasium.

The TI method incorporated a direct style of teaching including lectures, demonstrations, teacher questions and student questions. Students were given verbal instruction for 30-minute and they were able to ask questions. The instructor responded to questions asking for clarification, to repeat a verbal description, or to repeat a demonstration. Students were allowed to work alone or with a partner. There were 5-minute of introduction at the beginning of the period and the remaining time of approximately 10-minute was for questions and review.

Participants in the MCAI group were allowed to work independently or with a partner. The students were given 30-minute of computer time on a Pentium III computer. A multimedia program was developed for the purpose of this study, which was based on hypertext, graphics, animation, media and video. The MCAI program consisted of four topics, which corresponded precisely to theoretical work. There were 5-minute of introduction at the beginning of the period and the remaining time of approximately 10-minute was for questions and review. The instructor was present for organization and management supervision only. No verbal or visual reinforcement of any kind was offered by the instructor.

The CI group followed the same procedure, while implementing both the multimedia program and the traditional instruction. In the first three weeks the students participated with the traditional method group, and the remaining weeks with the MCAI method group. The theoretical session consisted of the same instruction and text information, which took place in the TI and the MCAI methods. Material for the three method groups was taken from a volleyball coaching textbook (American Sport Education Program, 2001) consisting of four topics: a) history, b) rules, court and player's position, c) skill fundamentals, and d) skill exercises.

At the end of the treatment, the knowledge test that previously served as a pre-test was given to students and the following day the attitude test was given to CI group as a post-test. One week later, the knowledge test procedure was repeated on the re-test to measure the level of retention in the selected setting skill. During the experiment, the participants in the three groups had no access to multimedia or to traditional learning environments beyond what was utilized as part of the experiment. All groups had the same learning conditions, such as topics and principles introduced in the treatments, and equal opportunities to achieve their learning outcomes.

Design

Due to practical limitation, a field experiment instead of a laboratory experiment was conducted to test the research questions. The experiment on knowledge test was a factorial design with teaching method groups (MCAI, TI and CI) and repeated measurements (pre-test, post-test and 1-week retention test) as independent variables, and knowledge learning as dependent variables. The experiment on attitude determination was a factorial design with CI teaching method group and post-test measurement as independent variables, and student's attitude as dependent variable.

RESULTS

Homogeneity of variance and Sphericity was verified by the Box's M test, the Levene's test and the Mauchly's test (Green & Salkind, 2007). Initial differences between the three groups for the mean knowledge scores were tested using one-way analysis of variance. An item analysis using the responses of the pilot study was conducted to determine the difficulty rating and index of discrimination. In determining the internal consistency of the knowledge test, the alpha reliability method was used. Two-way analysis of variance (ANOVA), with repeated measures on the last factor, were conducted to determine effect of method groups (MCAI, TI, CI) and measures (pre-test, post-test, re-test) on knowledge test. Paired-Samples t-tests were conducted to compare students' attitudes (CI group) towards the MCAI and TI teaching methods. Each variable was tested using an alpha level of significance .05. A listing of the results from the item analysis of the knowledge test in the pilot study can be found in table 2. Means and standard deviation for the MCAI, TI and the CI group in pre-test, post-test and re-test are presented on table 1, while results of each analysis are presented separately below.

Item Analysis

The pilot study knowledge test had a mean difficulty rating of 54%. When all items were analyzed, one question, or 6.6% of the items, had unacceptable difficulty rating values. The utilization of a difficulty rating criterion of between 10% and 90% resulted in 93.3% of the items yielding an acceptable level of difficulty. The pilot study knowledge test had a mean index of discrimination of .32. When all items were analyzed, two questions, or 13.2% of the items yielded an unacceptable index of discrimination values. The acceptable value for index of discrimination was .20 or higher. Acceptable index of discrimination values were observed for 86.6% of the items. Finally two more questions, or 13.2% of the items, had unacceptable index discrimination and difficulty rating values. As indicated by the information in table 2, five of the items (5, 6, 9, 14 & 15) were therefore deleted from the test for the main study.

Reliability Analysis

Reliability measures for knowledge test and attitude questionnaire were assessed. An alpha reliability coefficient .73 was computed based on the inter-item correlation coefficients of the pilot study knowledge test. While the Cronbach a coefficients of the attitude questionnaire was .83 and .93 for the two aspects of attitude: (1) positive and (2) negative, respectively, with all > .70. According to Green, & Salkind (2007), the reliability coefficient should be at least .70 for the test to be considered reliable. Thus, the determination was made that the pilot knowledge test and the attitude questionnaire were reliable measurement instruments.

Table 1. Means and standard deviations for pre-test, post-test and re-test scores of the three groups on knowledge test.

	Group	N	Mean	Std. Deviation
Knowledge Test 1st measure	TI	16	6,19	1,83
	MCAI	16	5,25	2,14
	CI	16	5,38	1,45
Knowledge Test 2st measure	TI	16	8,88	1,50
	MCAI	16	9,06	1,69
	CI	16	9,44	1,15
Knowledge Test 3st measure	TI	16	8,81	1,47
	MCAI	16	8,87	1,54
	CI	16	9,31	1,08

Table 2. Summary of Item Analysis for pilot study knowledge test.

Questions	Index of discrimination	Difficulty rating	Results
1	.40	40%	Retained
2	.28	55%	Retained
3	.25	59%	Retained
4	.35	47%	Retained
5	.69	18%	Eliminated
6	.15	73%	Eliminated
7	.28	55%	Retained
8	.49	34%	Retained
9	.11	78%	Eliminated
10	.25	59%	Retained
11	.28	55%	Retained
12	.62	22%	Retained
13	.59	25%	Retained
14	.01	96%	Eliminated
15	.01	96%	Eliminated

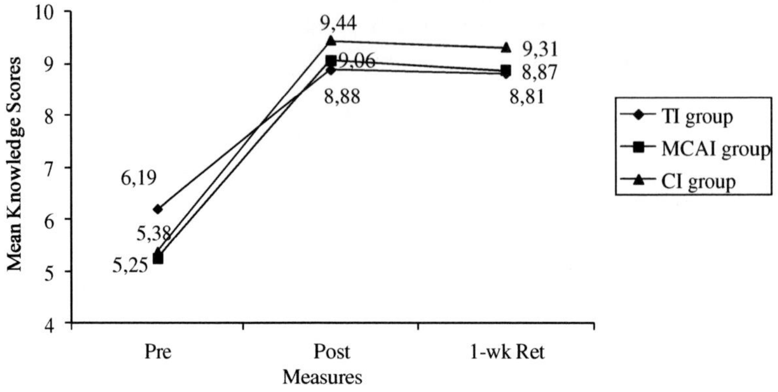

Figure 1. Performance of the three groups on all measures of the Knowledge Test.

Two-Way Repeated-Measures Analysis of Variance (ANOVA)

There were no significant initial differences between the three teaching method groups for the mean knowledge test scores, $F(2,45) = 1.234$, $p>.05$. A significant main effect was noted for the measurements, $F(1,45) = 92.623$, $p<.05$ but not for the group, $F(2,45) = .384$, $p>.05$, while the interaction measure X group was also not significant, $F(4,45) = 1.303$, $p>.05$.

Difference and repeated contrasts were conducted to follow up the significant measurements main effect. Differences in mean rating of knowledge test in TI group were significantly different between pre-test and post-test, $F(1,15)= 17.089$, $p<.05$ and between pre-test and re-test, $F(1,15)= 17.921$, $p<0,05$. Differences in mean rating of knowledge test in MCAI group were significantly different between pre-test and post-test, $F(1,15)= 25.948$, $p<.05$ and between pre-test and re-test, $F(1,15)= 30.190$, $p<.05$. Finally differences in mean rating of knowledge test in CI group were significantly different between pre-test and post-test, $F(1,15)= 65$, $p<.05$ and between pre-test and re-test, $F(1,15)= 78.502$, $p<.05$. As shown in Figure 1, the post-test and re-test knowledge scores were remarkably greater than pre-test knowledge scores for the three groups, while the difference between the post-test and re-test was not significant.

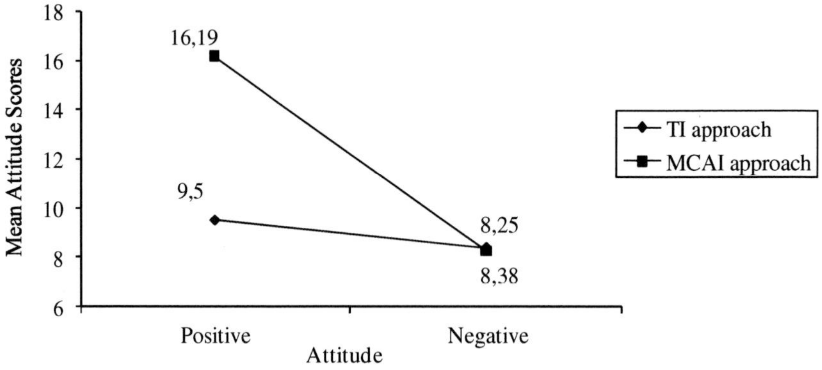

Figure 2. Profile plot of the positive and negative scores for each teaching approach.

Paired-Samples t Test Analysis

Paired-samples t test were conducted to evaluate whether CI group students were more favourable with TI or MCAI teaching approaches. The results indicated that the mean positive attitude (M=16.19, SD=6.54) was significantly greater than the mean negative attitude (M=8.25, SD=3.21) for the MCAI teaching approach $t(15)=5.277$, $p<.01$. No significant difference was found between the mean positive attitude (M=9.5, SD=4.75) and the mean negative attitude (M=8.38, SD=7.68) for TI teaching approach $t(15)=.649$, $p>.05$. Also, significant difference was found between the mean scores of the MCAI teaching approach (M=16.19, SD=6.54) and the mean scores of the TI teaching approach (M=9.5, SD=4.75) for the positive aspect of the questionnaire $t(15)=3.105$, $p<.01$. On the contrary, no significant difference was found between the mean scores of the MCAI teaching approach (M=8.25, SD=3.21) and the mean scores of the TI teaching approach (M=8.38, SD=7.68) for the negative aspect of the questionnaire $t(15)=.093$, $p>.05$. As shown in Figure 2, the CI group students were more favourable with MCAI teaching approach than TI teaching approach.

CONCLUSION

The present study was designed to examine differences that may occur when individuals learn a motor skill under different instructional teaching methods and replicated previous findings by showing differential performance dependent on teaching methods. With regard to the knowledge test, all groups improved their cognitive learning in volleyball setting skill, after instruction. Post-test results indicated no significant differences between the groups concerning the written test. Nevertheless, the attitude test scores of the CI group were more favourable to MCAI method than TI method. Retention test results showed that groups retained the knowledge acquisition. However, the combine method of instruction tended to be the most effective on cognitive learning.

The significant improvements in knowledge test scores from the pre-test to the post-test for the TI group, the MCAI group and the CI group while not having significant differences between the groups on the learning variable are consistent with the results reported in the literature. Wiksten, Spanjer, & LaMaster (2002) did not find a significant learning difference between computer-assisted instruction and the traditional method of instruction in an introductory athletic training laboratory class on either the written or practical examination. Vernadakis et al. (2002) developed a multimedia application for the instruction of the setting skill in volleyball and reported that the study provided no evidence to indicate that the multimedia computer-assisted instruction application contributed significantly more to student learning when compared to the control group which had been instructed by traditional means. In earlier studies comparing the impact of MCAI and TI method, Adams, et al. (1991) and Kerns (1989) found no significant differences in scores of tests in golf and tennis rules.

On the other hand, there are previous studies in physical education supporting the view that computer assisted instruction is consistently superior to the traditional forms of instruction (Antoniou et al., 2006; Si-min Li and Jin-hai Sun 2008) while other evidence is contradictory. From a recent multimedia study on learning rule violations in basketball, Antoniou et al. (2003) found that university physical education students receiving lecture instructions performed significantly better than the MCAI group. In other study, Siskos, Antoniou, Papaioannou, and Laparidis (2005) found that MCAI was superior to traditional classroom teaching in passing knowledge in health-related fitness and nutrition subjects.

The lack of significant differences from the post-test to the re-test for the three instructional method groups is inconsistent with the meta-analytic study (Cotton, 1991) which integrated the findings of 59 independent evaluations of computer-assisted instruction during the 1980s and reported that CAI as a supplement to traditional teacher-directed instruction produces maintenance scores superior to those obtained with traditional instruction alone. Research is inconclusive regarding the comparative effectiveness of conventional instruction alone and CAI alone. Antoniou et al. (2003), for example, found that traditional lecture instruction was superior to MCAI in the retention of basketball rule violations knowledge. On the contrary, Yildirim, Ozden, & Aksu, (2001) reported that multimedia applications increase the ability of information maintenance. Also, Vernadakis, Avgerinos, Zetou, Giannousi & Kioumourtzoglou, (2006) found no significant differences between post-test and re-test for the three instructional methods (TI MCAI & CI) in the long jump event knowledge test.

However, Mayer (2001) reported that well-designed multimedia applications increase the ability of information maintenance. In the current study the evaluation of the learning maintenance, showed that the three instructional method groups mean scores did decrease slightly from the post-test to the re-test, however the difference was not significant. The time period between the administration of the post-test and the re-test was short, 7 days. While the re-test mean scores indicated that learning may have been affected differently during this period for the tree instructional groups, long term retention differences would be purely speculative.

The attitude survey showed that the MCAI teaching method had a stronger positive effect on students' attitude, than the TI teaching method. Students did state that learning about volleyball setting skill on the multimedia application made the subject matter more interesting, which could imply that this method of teaching can be used effectively. Students rated the activities as intelligent, meaningful, and pleasant for learning and remembering, and they particularly valued the interactive nature of the learning experience and the visual nature of the presentation. They seemed to understand and comprehend the material better. In light of this, these findings reaffirm previous studies that concluded that the use of multimedia computer-based instructional programs improve students' attitude scores as compared to the use of traditional or regular methods of instruction (Astleitner and Wiesner, 2004; Goran and Reynolds, 2005; Kiboss and Ogunniyi, 2003; Liao, 1999).

However, Bernard, et al. (2004) in a comprehensive meta-analysis of 232 empirical studies conducted between 1985 and 2002, concluded that courses reporting high levels of student interest also tended to report lower levels of achievement. They also concluded that end of course measures of interest tended to be negatively correlated with end of course achievement. Thus, as achievement increased in multimedia studies, student interest and satisfaction decreased. They conclude that interest satisfaction may not indicate success but the opposite, since students may spend less effort learning, especially when they choose between multimedia education and regular courses for convenience purposes.

Evaluating the outcomes of the present research study, greater consideration needs to be given to those factors that might strongly affect students' learning. First, students were from one middle school of Thessaloniki. A larger and more diverse sample would provide a more stringent test for cognitive learning on a MCAI program. Additionally, the results reported in this study are based on a single MCAI program. This is a case-specificity problem. It is possible that a different type of MCAI package covering different content would yield different results.

Secondly, the age of students might be critical when it comes to learning independently. Since the participating students were all around 13 years of age when this study was conducted, they might not possess the learning skills that are needed in order to work independently using individual computers. Besides, research has also found that first-time users of computers are often overwhelmed by the vast amount of materials and information that can be presented by multimedia courseware (Renshaw and Taylor, 2000). These types of differences between the three groups, therefore, might be reflected in the students' post-test and 1-week retention knowledge test scores.

Thirdly, no attempt was made to control possible differences in computer skills and multimedia experiences of the students or the effective learning time of the students' real engagement in multimedia learning. If these limitations have been controlled and the effective learning time had lasted longer, the researchers might have reported more precise results for the effectiveness of MCAI, TI, and CI methods on cognitive learning of the volleyball setting skill. For those reasons, further research may be needed to replicate this study.

Those limitations of the research learning environment may have significantly affected the experimental groups' ability to learn and to retain the knowledge acquisition of the volleyball setting skill. However, it would be difficult to be certain, that the MCAI group would have been more successful than TI and CI groups on cognitive learning if the above limitations could have been eliminated. In that sense, these results indicate that students can be taught through the use of multiple effective teaching techniques. Multimedia programs have been generally successful especially when it has been used in connection with regular classroom instruction (Vernadakis, Antoniou, Zetou & Kioumourtzoglou, 2004).

In conclusion, multimedia programs can be utilized to enhance the effectiveness of teaching strategies or techniques in physical education classes. Computers can be used for the teaching of the cognitive aspects of sports such as rules and scoring procedures, and to allow teachers to have more time to spend with students' motor skills. However, these conclusions are limited for students aged 12 – 14 years old. More studies should be conducted to investigate the effect of MCAI in different ages and for various sport activities. Also, it is critical to continue researching into how students learn in different technological environments. Subsequent research with more attention to software design and validation may begin to show more consistent results in the use of MCAI software. These results lend support to the supplemental role of MCAI to more traditional instructional methods in the context presented in this investigation.

REFERENCES

Adams, T., Kandt, G., Throgmartin, D. & Waldrop, P. (1991). Computer – Assisted Instruction vs Lecture Methods in Teaching the Rules of Golf. *Physical Educator, 48(3)*, 146-150.

Akhtar, A. (2003). *A study of interactive media for deaf learners in post 16 education.* Paper presented at the Instructional Technology and Education of the Deaf Symposium,

Rochester, NY. Retrieved July 13, 2009, from http://www.rit.edu/~techsym/papers/2003/W3A.pdf

American Sport Education Program (2001). *Coaching youth volleyball* (3rd ed). Human Kinetics, Champaign IL.

Antoniou, P., Derri, V., Kioumourtzoglou, E. & Mouroutsos, S. (2003). Applying multimedia computer-assisted instruction to enhance physical education students' knowledge of basketball rules. *European Journal of Physical Education., 8(1)*, 78-90.

Antoniou, P., Moulelis, E., Siskos, A. & Tsamourtzis, E. (2006). Multimedia: an instructional tool in the teaching process of alpine ski. *Current Developments in Technology-Assisted Education.* (941-945). Retrieved July 17, 2009, from http://www.rit.edu/~techsym/papers/2003/W3A.pdfhttp://www.formatex.org/micte2006/pdf/941-945.pdf.

Antoniou, P., Theodorakis, Y. & Kioumourtzoglou, E. (1999). Teaching differences of the relaxation technique between traditional teaching and the one by means of computer. *Exercise & Society., 38*, 73-83.

Astleitner, H. & Wiesner, C. (2004). An Integrated Model of Multimedia Learning and Motivation. *Journal of Educational Multimedia and Hypermedia., 13(1)*, 3-21.

Bangert-Drowns, R. L. (1993). The word processor as an instructional tool: A Meta-analysis of word processing in writing instruction. *Review of Educational Research., 63*, 63-93.

Bernard, R., Abrami, P., Lou, Y., Borokhovski, E., Wade, A., Wozney, L., Wallet, P. A., Fiset, M. & Binru Huang. (2004). How Does Distance Education Compare With Classroom Instruction? A Meta-Analysis of the Empirical Literature. *Review of Educational Research., 74(3)*, 379-439.

Blok, H., Oostdam, R., Otter, M. E. & Overmaat, M. (2002). Computer-Assisted instruction in support of beginning reading instruction: A review. *Review of Educational Research, 72(1)*, 101-130.

Cotton, K. (1991a). *Computer-assisted instruction* (SIRS Close-up No. 10). Portland, OR: Northwest Regional Educational Laboratory. Retrieved July 12, 2009, from www.nwrel.org/scpd/sirs/5/cu10.html

Everhart, B., Harshaw, C., Everhart, B., Kernodle, M. & Stubblefield, E. (2002). Multimedia Software's Effects on High School Physical Education Students' Fitness Patterns. *Physical Educator., 59(3)*, 151-157.

Goran, M. & Reynolds, K. (2005). Interactive Multimedia for Promoting Physical Activity (IMPACT) in Children. *Obesity research., 13(4)*, 762-771.

Green, B. S. & Salkind, J. N. (2007). *Using SPSS for Windows and Macintosh* (5th ed.). New Jersey: Prentice Hall.

Kerns, M. (1989). The effectiveness of computer-assisted instruction in teaching tennis rules and strategies. *Journal of Teaching in Physical Education., 8(2)*, 123-130.

Kiboss, J. K. & Ogunniyi. M. B. (2003). Influence of a computer-based intervention on students' conceptions of measurement in secondary school physics in Kenya. *Themes in Education, 4(2)*, 203-217.

Kulik, C. C. & Kulik, J. A. (1991). Effectiveness of computer-based instruction: An updated analysis. *Computers in Human Behavior., 71*, 75-94.

Liao, Y. C. (1999). Effects of hypermedia on students' achievement: A meta-analysis. *Journal of Educational Multimedia and Hypermedia., 8*, 255-277.

Mayer, R. E. (2001). *Multimedia learning*. New York: Cambridge University Press.

McGee, R. & Farrow, A. (1987). *Test questions for physical education activities*. Champaign, IL: Human Kinetics.

Renshaw, C. E. & Taylor, H. A. (2000). The educational effectiveness of computer-based instruction. *Computers and Geosciences.*, *26*, 677-682.

Simatos, A. (2000). *Technology and Education*. Athens: Patakis Press.

Si-min Li, & Jin-hai Sun (2008). Experimental Research on Multimedia Teaching for Sports Aerobics. International Conference on MultiMedia and Information Technology. (pp. 711-714). Retrieved July 21, 2009, from http://www2.computer.org/portal/web/csdl/doi/10.1109/MMIT.2008.48.

Siskos, A., Antoniou, P., Papaioannou, A. & Laparidis K. (2005). Effects of multimedia computer-assisted instruction (MCAI) on academic achievement in physical education of Greek primary students. *Interactive Educational Multimedia.*, *10*, 61-77.

Vernadakis, N., Antoniou, P., Zetou, E. & Kioumourtzoglou, E. (2004). Comparison of Three Different Instructional Methods on Teaching the Skill of Shooting in Basketball. *Journal of Human Movement Studies.*, *46*, 421-440.

Vernadakis, N., Avgerinos, A., Zetou, E., Giannousi, M. & Kioumourtzoglou, E. (2006). Comparison of Multimedia Computer Assisted Instruction, Traditional Instruction and Combined Instruction on Learning the Skill of Long Jump. *International Journal of Computer Science in Sport.*, *5(1)*, 17-32.

Vernadakis, N., Zetou, E., Antoniou, P. & Kioumourtzoglou, E. (2002). The Effectiveness of Computer – Assisted Instruction on Teaching the Skill of Setting in Volleyball. *Journal of Human Movement Studies.*, *43*, 151-164.

Wiemeyer, J. (2003). Learning with Multimedia - More Promise than Practice? *International Journal of Computer Science in Sport.*, *2(1)*, 102-116.

Wiksten, L. D., Spanjer, J. & LaMaster, K. (2002). Effective Use of Multimedia Technology in Athletic Training Education. *Journal of Athletic Training.*, *37(4)*, S-213-S-219.

Wilkinson, C., Hillier, R., Padfield, G. & Harrison, J. (1999). The effects of volleyball software on female junior high school students' volleyball performance. *Physical Educator.*, *56(4)*, 202-209.

Yildirim, Z., Ozden, M. Y. & Aksu, M. (2001). Comparison of Hypermedia Learning and Traditional Instruction on Knowledge Acquisition and Retention. *The Journal of educational Research.*, *94(4)*, 207-214.

APPENDIX

Appendix A: Knowledge test

Circle or underline the correct answer

1. How should the player's body be positioned in relation to the ball in hitting an overhand pass?

   a. the approaching ball should be in line with the arm and shoulder of the player's dominant side

b. the approaching ball should be in line with the arm and shoulder of the player's right side
c. the approaching ball should be in line with the midline of the player's body *
d. the approaching ball should be in line with the arm and shoulder opposite the intended flight of the ball

2. What parts of the hands should contact the ball in hitting the overhand pass?

a. the heels of the hands, finger pads, and thumbs
b. the finger pads and thumbs *
c. the palms of the hands, finger pads, and thumbs
d. the thumbs and finger pads of the first two fingers

3. Where should contact with the ball be made in hitting an overhand pass?

a. 6 inches to a foot directly above the head
b. 6 to 8 inches off the forehead *
c. 6 to 8 inches out from the nose
d. 6 inches to a foot out from the chest

4. What is the position of the elbows at the beginning of an overhand pass?

a. in close to the body at the waist
b. in close to the body at the chest *
c. away from the body at the waist level
d. away from the body at the shoulder level

5. What is the most important factor to remember in setting?

a. to move quickly to get the ball
b. to push the ball out and up
c. to position oneself under the ball
d. to place the ball in the proper spiking zone *

6. What is the approximate size of a volleyball court?

a. 30 by 30 feet
b. 30 by 60 feet *
c. 60 by 60 feet
d. 60 by 90 feet

7. How many players may be on a team including substitutes?

a. 8
b. 10
c. 12 *

d. 15

8. If a ball is not hit on a block, how many times may a team contact the ball before it must go over the net into the opponent's court?

   a. one
   b. two
   c. three *
   d. four

9. How many substitutions is a team allowed each game in international competition?

   a. 4
   b. 6 *
   c. 10
   d. 12

10. How many games must be played to make a match in international play?

   a. two
   b. best two out of three
   c. best three out of five *
   d. best four out of seven

APPENDIX B: POST-TEST ATTITUDE SCALE

Express your feeling with the following statements on a 0-5 scale where 0=weak and 5=powerful by comparing TI and MCAI teaching methods.

TI	←									→	MCAI
good	5	4	3	2	1	0	1	2	3	4	5	good
bad	5	4	3	2	1	0	1	2	3	4	5	bad
foolish	5	4	3	2	1	0	1	2	3	4	5	foolish
intelligent	5	4	3	2	1	0	1	2	3	4	5	intelligent
meaningful	5	4	3	2	1	0	1	2	3	4	5	meaningful
meaningless	5	4	3	2	1	0	1	2	3	4	5	meaningless
unattractive	5	4	3	2	1	0	1	2	3	4	5	unattractive
attractive	5	4	3	2	1	0	1	2	3	4	5	attractive
ugly	5	4	3	2	1	0	1	2	3	4	5	ugly
beautiful	5	4	3	2	1	0	1	2	3	4	5	beautiful
unpleasant	5	4	3	2	1	0	1	2	3	4	5	unpleasant
pleasant	5	4	3	2	1	0	1	2	3	4	5	pleasant

In: Computer-Assisted Teaching: New Developments ISBN: 978-1-60876-855-4
Editors: B. A. Morris, G. M. Ferguson, pp. 151-165

Chapter 7

NEW FORMS OF AND TOOLS FOR COOPERATIVE LEARNING WITH SOCIAL SOFTWARE IN HIGHER EDUCATION

Sandra Schaffert[1*] and Martin Ebner[2**]
[1]Salzburg Research Forschungsgesellschaft, Salzburg, Austria,
[2]Social Learning Department, Computer and Information Services, Graz University of Technology, Austria

ABSTRACT

Since the new generation of Internet technology, called Web 2.0, has been introduced, a change of how users are dealing with the World Wide Web has been get into going. If access to the Web is available, today nearly anyone can actively participate and communicate online. Of course this recent evolution of the Web influences also the field of education. Former e-learning was mainly characterized by the use of content offered within learning management systems. Nowadays so called "Social Software" enables new possibilities and didactical approaches. In this chapter we give a short overview of how Social Software can support cooperative learning and how new technologies can enhance higher education in a meaningful new way. After a short introduction to the basics of cooperative learning different Social Software applications are classified and described. Practical examples are presented to show the general usage. In the end we conclude that these technologies have great impact on teaching and learning, as it will help to enhance education at universities.

* Corresponding author: sandra.schaffert@salzburgresearch.at
** martin.ebner@tugraz.at

The Concept of Cooperative Learning and its Technological Support

According to the currently dominating social constructivism theory, learning is not a passive, receptive process, but an active and constructive one, where "the others" play an important role. Learning is an active process on the part of the learner, where knowledge and understanding is constructed by the learner (Soloway & Bielaczyc, 1996; Holzinger, 2002) and moreover it must strongly be considered as social process: learning proceeds by and through communication (Preece, 2000). Vygotsky (Vygotsky, 1978) mentioned in his theory about interaction and learning that problem solving and similar approaches occur under guidance or in collaboration with capable peers. With other words it can be summarized that education must be seen as collaborative process, which is proceeded through communication.

Among other aspects, the development of a learning environment should always include a possibility for learners to reflect and compare the ideas and experiences of others (cf. Gräsel, Bruhn, Mandl & Fischer, 1997). Cooperative and collaborative learning arrangement, for example project and group work as well as group discussion are therefore seen as adequate measures to support knowledge and competence development. A review on 25 years of cooperative learning displays that the exchange of different experiences and concepts of peers helps to reflect on own (mis-) conceptions and therefore is seen as crucial from the perspective of developmental psychology (Slavin, 1997, 10).

Nevertheless, in the early years of computer supported learning the social aspects of learning had been overseen or ignored. Research and practice concentrated on the possibilities of programmed learning, implementation of instructional design and artificial intelligence. Since the beginning of the 1990ies, with Internet services such as electronic mail, Usenet and the World Wide Web, the role of peers and tutors in computer supported learning environments gained more and more attention. First research on computer-supported collaborative learning (CSCL) analysed for example the collaboration of pupils in schools in different cities corresponded via Quickmail and their development of common project work (Campione, Brown & Jay, 1992).

Current "Social Software" technologies and applications are characterised by their high potential of bringing people together through facilitating communication and collaboration. We use the term "Social Software" as the sum of all old and new forms of tools and applications that can be or are ordinary used for communication and collaboration. Due to the success of the online encyclopaedia Wikipedia, the Wiki technology (now) has become a famous example of an invention of a very simple way to create and edit Webpages in a collaborative way.

Social Software can be distinguished concerning their main purposes

- Social presence and communication including discussion forums, Web chats, (micro-) blogging, (micro-) podcasting, and live streaming;
- Collaborative development including tools that allow a collaborative work and development as the Wiki technology; or
- Collaborative enrichment of content such as social bookmarking, social tagging, and rating.

There are currently a lot of platforms on the Web that combine all of these three aspects, or bring several applications due to development of mash-up standards and technologies together: For example, social networking platforms as LinkedIn[1] or Facebook[2] support a huge amount of different ways for social interaction using the Web.

Social Software is potentially usable in cooperative learning settings. Nevertheless, this requires certain equipment and competencies of learners and teachers as well as the experiences and research on their usage and usability for learning and teaching is often hardly elaborated.In this chapter we describe how current Social Software can be used and already is in use in higher education within different forms of cooperative learning arrangements. For this, we have included new tools that seem especially usable in cooperative learning settings. Some of them were taken from some Web's "best of"-lists of tools[8] for learning and teaching or were found reading current publications about Social Software for learning. Especially for European projects, which have been co-financed by the European Commission, the Prolearn[3]and the I-Camp[4] projects are to be named in this context.

PROTOTYPICAL SETTINGS OF COOPERATIVE LEARNING WITH SOCIAL SOFTWARE

Before we go on with a description of new tools and their usage for cooperative learning we want to clarify that nowadays there is no uniform or single standardized setting of cooperative learning and teaching in higher education; the possibilities are multiple and diverse. From a practical and didactical perspective there are four different prototypical settings for computer supported cooperative learning that can currently be distinguished for formally organised learning:

1. Distance learning setting
   The learners are distributed and do not meet in reality before and while learning and working together. The online communication is not always, but often asynchronous;
2. Blended learning setting
   The learners meet in reality and additionally online but normally not parallel. The online communication usually is asynchronous;
3. Classroom group work setting - with 10 to 25 learners
   Every learner additionally and parallel to "real" communication participates through networked computers or mobile phones on the group interaction
4. Lecture hall learning setting
   The learners use networked computers or mobile phones to facilitate interaction and feedback loops in big groups of more than 40 people parallel to a (interactive) lecture.

---

[1] http://www.linkedin.com (last view: August 2009)

[2] http://facebook.com/ (last view: August 2009)

[3] http://www.prolearn-project.org/ (last view: August 2009)

[4] http://www.icamp-project.org/ (last view: August 2009)

Besides these settings, which are institutionally organised and known as "formal" learning settings, cooperative learning also occurs in informal, self-organised or incidental learning while being online. Learning can be initiated and supported by being actively involved in a community of learners or interests; it can also be the result of (or reflection on) things seen or done in collaboration with others, more loosely connected persons in the Web, such as unknown or anonymous editors of a Wikipedia page. (Informal) online learning communities and groups are using different tools or platforms in the Web. Nevertheless, there are some Websites that are explicitly developed for cooperative learning: Grockit[5] develops a learning game; Livemocha[6] and Busuu.com[7] are language learning platforms where every learner serves as a teacher in her/his native language.

In the following, we concentrate on formally arranged cooperative learning settings in the field of higher education. In general, the following examples and descriptions cannot be seen as usual practice in higher education nor for current students' abilities and experiences within the Web (cf. Ebner Schiefner & Nagler, 2008a; Jadin & Zöserl, 2009; Nagler & Ebner, 2009). The majority of these examples are innovative and not common experiences in current higher education. In the following, we introduce typical applications and (first) experiences in higher education according to the above mentioned three types of characteristics of Social Software, which are communication, collaboration and collaborative enrichment. The focus of this contribution is not to describe these tools in detail, but to give some examples how they can be used for different settings of cooperative learning in higher education.

SOCIAL PRESENCE AND COMMUNICATION

The first sort of applications, which can be used for cooperative learning settings, allows to communicate or (at least) to indicate presence, for example a current status or mood. In 1968, Licklider (Licklider & Taylor, 1968) pointed out that "men will be able to communicate more effectively through a machine than face to face" and without having any idea of today's World Wide Web, he noticed the growing importance of communication through a large network of connected devices. Especially for online environments where face-to-face interaction is replaced by status updates and text-based activities social presence became a new dimension; as DuVall (DuVall et al., 2007) stated that the learner's satisfaction depends on it. Some research work clearly mentioned that social presence affects the support of critical thinking and engages learners for online collaboration (Garrison, 2003).

Of course communication is possible with the help of various tools, applications and programs. This includes discussion forums and Web chats (e.g. Tinychat[8]), but also distributed communication channels as Weblogs or micro-blogging tools. (Micro-) blogging, including text posts but also podcasts, is often described as personal online journal. Nevertheless it allows learners to easily interact by commenting or interlinking. Besides these Web tools, instant messaging services allow a sort of private, real time chat room for two or

---

[5] http://grockit.com/ (last view: August 2009)

[6] http://www.livemocha.com/ (last view: August 2009)

[7] http://www.busuu.com/de (last view: August 2009)

[8] http://tinychat.com/ (last view: August 2009)

more people, e.g. Skype[9]. Conference systems additionally allow presenting the own screen or presentation slides.

Classic Communication Tools

There are several possibilities to use these communication tools for cooperative learning in higher education. In general, discussion forums, chats and also, but perhaps to a limited extent, instant messaging are quite common tools in distance education and blended learning settings, to distribute and exchange information.

All these communication tools can support cooperative learning, for example for a group work: They can be used to clarify the task and to organise the group or simple to exchange information. Also in face-to-face learning situations, instant messaging is a very easy way to exchange or distribute e.g. files or hyperlinks with further information (Dulik, 2009). Unlike electronic mail most of these communication tools as instant messaging or chats allow to see the current status of other learners, whether they are online and available for a short chat. Some of them also provide information about the mood of the users online.

Especially discussion forums and newsgroups have a long tradition in online learning environments for asynchronous communication. Some research work gave insights into how learning communities can be motivated to exchange their thoughts through these communication channels (Salmon, 2002; Wenger, 2002). Discussion forums are mainly used in higher education to address to problems, to exchange information regarding the lecture topic and to provide thoughts and understandings. A further didactical approach is to use newsgroups for online roleplays and peer-review processes among learners (Bell, 2001). Beside that so-called thematic uploads (discussion to a specific topic, file or sketch) as well as enhanced FAQ lists (discussion directly on the question) were tested at Graz University of Technology (Ebner, Scerbakov & Maurer, 2005) with the aim to support discussions at their point of occurring.

Micro-Blogging

Micro-blogging can be seen as latest variant of blogging where messages are posted more or less instantly and users are updating their activities, thoughts, everyday experiences, moods and feelings constantly. Living in an online stream becomes reality. Templeton (2008) defines micro-blogging as "a small-scale form of blogging, generally made up of short succinct messages, used by both consumers and business to share news, post status updated and carry on conversations." Currently an enormous trend towards integrating micro-blogging activities is observable. Facebook[10], Friendfeed[11] and Twitter[12] are the most famous applications for exchange small pieces of data amongst a huge worldwide community. Different research works pointed out that micro-blogging must be seen as a new form of communication

---

[9] http://skype.com/ (last view: August 2009)

[10] http://www.facebook.com (last visit: August 2009)

[11] http://friendfeed.com/ (last visit: August 2009)

[12] http://twitter.com/ (last visit: August 2009)

(McFedries, 2007) to exchange mainly four types of messages (Java et al., 2007) – daily routines, simple conversation, sharing information or reporting news.

Similar to other studies about the use of Short Message Services (SMS) for learning purposes (DuVall, 2007) micro-blogging can serve in higher education for the following reasons:

- Exchange of information, thoughts, ideas amongst people of same interest (Ebner & Schiefner, 2008b)
- To enhance the classroom discussion about specific topics and for recording research activities on the Web or in general (Ebner & Maurer, 2008c)
- To reporting live from events, lectures and presentations and usage as an (informal) back channel (Ebner, 2009a; Reinhardt et al., 2009)

Weblogs

Also Weblogs can be seen as a further communication tool, because they can be easily interlinked and allow discussion of single posting. As Barlett-Bragg (Barlett-Bragg, 2003) mentioned Weblogs can be used in many different ways: as a community blog, to publish student writings, to publish field notices, to have a journal for your professional worked, to publish your own meaning, as instrument to reflect your research and as learning journal. Nevertheless, this means that the main focus of their usage in cooperative learning is not a group discussion or a group work. Weblogs in higher education are normally used to reflect own learning content and processes. Keeping an e-portfolio work learners are asked to define learning goals, steps and tasks as well as to document and reflect their learning using special e-portfolio software or a Weblog. Following the three principles of the Blogging Theory (Schiefner & Ebner, 2008) – individuality, collectivity and community – a crucial factor of a successful e-portfolio is to exchange reflections with people of same interests. The importance of the peers is essential for ones e-portfolio. In this context Weblogs and their possibilities to comment and interlink, to reflect and discuss the other's contribution gets of higher interest.

Communication Possibilities in Mass Education

One of the latest trends in the area of e-learning is the use of mobile devices for learning and teaching purposes called mobile learning (m-learning). Together with the dramatic increase of powerful mobile devices (e.g. iPhone) and the availability of Wi-Fi networks[13] mobiles can be used as an integral part of learning as well as teaching activities (Norris & Soloway, 2004) even in traditional mass education in higher education.

To overcome well known problems of big lecture halls (Anderson et al., 2003) – missing feedback, fear to ask during a lecture and the only-one-speaker syndrome – different research studies address to bring more interactivity to the classroom and with that opportunities for

[13] http://en.wikipedia.org/wiki/Wi-Fi (last visit: August 2009)

cooperative learning. In former years most projects concentrate on the use of Personal Digital Assistants (PDAs) in combination with servers.

Table 1 Comparison of traditional interactive tools (Ebner, 2009b)

System	Interaction type (Student - lecturer)	Used devices	Environment
Classtalk[14]	Lecturer is able to force questions to students' devices	PDA	Stand alone
ClassInHand[15]	Presentation control application; quiz feature	PDA	Stand alone
ConcertStudeo[16]	Multiple-choice quizzes, queries, brainstorming sessions	Electronic blackboard and handheld devices	Stand alone
CFS (Anderson et al. 2003)	Online feedback (annotations to presentations)	Notebooks	Stand alone
ActiveClass (Ratto et al. 2003)	Feedback, Quizzes	PDAs	Stand alone
Mobile Notes (Bollen et al. 2006)	Adding notes	PDAs	Stand alone

Table 1 (Ebner, 2009b) gives a short overview about existing systems. The column "interaction type" describes the interaction type between lecturers and students. Mainly students are able to ask questions, give feedbacks, vote or fill out quizzes by using PDAs or special mobile devices. A further project at Vienna University of Technology focuses on the technology and the ideas of collaboration with Social Software. Using the push-technology students can follow the current presentation of the lecturers on their own devices just in time (slides are automatically changed on learner's devices) and each learner can share his/her notes collaboratively online (Purgarthofer & Reinthaler, 2008). A further report pointed out that also micro-blogging from mobile devices can be used to comment, ask questions or even discuss presenter's slides directly and just in time to improve teacher-student interaction in lecture halls (Ebner, 2009b). Mobile learning applications can be seen as additional channels to support the communication flow in big lecture hall learning.

Further Tools

Finally there are some hints to further communication possibilities to arrange and coordinate learning activities. For example, students can use tools to sort their notes digitally (e.g. postit[17]) and share them with their colleagues online. Mobile access allows doing it just in time and anywhere. But not only notes also arranging groups, projects or learning meetings

[14] http://www.bedu.com/ (last visit: August 2009)

[15] http://classinhand.wfu.edu/ (last visit: August 2009)

[16] http://www.ipsi.fraunhofer.de/concert/index_en.shtml?projects/past_projects/studeo (last visit August 2009)

[17] http://www.lokalguide.com/postit/ (last visit: August 2009)

can be arranged by using appropriate Web 2.0 tools (e.g. Doodle[18]). In the end the whole learning process can be planned, recorded and shared (e.g. PlanItEasy[19]) even with modern mobile phones (e.g. EduCate[20]).

COLLABORATIVE DEVELOPMENT

Web 2.0 technologies made collaborative work much easier as known before. Users can contribute to the World Wide Web without any knowledge of any programming language, even HTML scripting. Tim O'Reilly (O'Reilly, 2005), who named the Web 2.0 for the very first time, summarized this step towards a World Wide Web where "users are generating the content". From the learning and teaching perspective beyond communicating the exchange of digital data as well as any information gets possible. In the following, the most interesting collaboration applications are described and how they can be used in higher education scenarios.

Wiki

The concept of Wikis, which was introduced by Bo Leuf and Ward Curringham in 1995 (Leuf & Curringham, 2001), bases on the idea that users create, edit, revise, extend or link articles within an online platform. The most famous used Wiki system is the well-known online encyclopaedia Wikipedia[21], where thousands contributors voluntarily write on creating the world's largest open content project. Concerning the use in higher education, Wikis are appropriate tools for collaboration amongst a group of learners (Jaksch et al., 2008) or even classrooms (Ebner et al., 2008d). Furthermore learners can create their own learning and collaboration space for documentation. Very new research work also combines the method of geo-tagging (pictures enhanced by global geo-coordinates) and Wiki technology to allow field work to display locations in real-time using additional services like Google Maps (Safran et al., 2009).

Sharing Documents

A very important aspect in online collaboration is to exchange files and documents between users. Each learner should have access to the latest version of the collaborative work. Therefore different online tools help to organize and manage this purpose. On the one side, there are applications that allow similar to the Wiki principle to create and edit the document online and to export it at last to the appropriate format (e.g. Google Docs[22], Writely). On the other side, there are also numerous online services that are providing Web space for placing

---

[18] http://www.doodle.com (last visit: August 2009)
[19] http://www.planiteasy.de/ (last visit: August 2009)
[20] http://www.ikonstrukt.com/ (last visit: August 2009)
[21] http://en.wikipedia.org/wiki/Main_Page (last visit: August 2009)
[22] http://docs.google.com (last visit: August 2009)

documents that can be accessed by small desktop applications or within common local file-management systems (e.g. Dropbox[23]).

In any case the possibility of distributed collaborative working will be one of the next major steps in the online world called "Cloud Computing". It can be expected that participating in different services on the Web and collaborating between learners and learners as well as learners and teachers will increase enormously.

Weblogs

Regarding the use of Weblogs for collaborative development in higher education there are mainly two possibilities:

- Aggregation of Weblogs
  Each learner conducts his/her own Weblog for reflection, documentation and exchange and hyperlinks to other contributions using methods like trackback and pingback (Helen & Wagner, 2006). With other words a learner network is built, a blogosphere for learning and teaching purposes.
- Cooperate Blogging
  In this case all learners are using the same Weblog for their contributions. This form is used when users are writing about the same topic over a period of time. Examples show that the content of the lecturers becomes more student-centred and student-driven. Lecturers have to moderate the whole process (Ebner & Maurer, 2007).

Further Tools

Further interesting applications are concerning the possibility to collaborate in real-time. For example, learners can develop some sketches collaborative just-in-time by using online whiteboards (e.g. skrbl[24]), write a short essay instantly (e.g. etherpad[25]) or create mind maps (e.g. mindmeister[26]). Even the construction of timelines[27] or concepts and their relationships (e.g. Conzilla[28]) (Palmer & Naeva, 2005) can be realized collaborative online. Klamma et al. (2007) pointed out that all these features are leading to a collaborative adaptive learning platform based on semantic technologies. They improve the existing learning environments by fulfilling the users' needs by supporting their highly challenging learning tasks.

---

[23] http://www.getdropbox.com (last visit: August 2009)

[24] http://www.skrbl.com/ (last visit: August 2009)

[25] http://etherpad.com/ (last visit: August 2009)

[26] http://www.mindmeister.com/de/home (last visit: August 2009)

[27] http://www.xtimeline.com (last visit: August 2009)

[28] http://www.conzilla.org (last visit: August 2009)

COLLABORATIVE ENRICHMENT OF CONTENT

Beside the communication and the collaborative development of content the third main purpose of Social Software concerns the collaborative enrichment of content. By adding keywords to different kind of content (tagging) users are able to retrieve and find their stored items easier. Tagging helps to categorize within a big database and to provide different kind of recommendations. Therefore, collaborative enrichment of content can be the core of a cooperative learning setting, but is mostly an indirectly consequence of common usage of the tools.

Social Bookmarking

Social bookmarking describes the possibility to store bookmarks online and to tag them with keywords. Del.icio.us[29], the most famous application further allows to share and to comment the bookmarked items among their users. In higher education social bookmarking helps to share items of research and online resources within a huge group of learners. Within a very short time frame (Ebner, 2009c) the search results are provided automatically to the whole learning community.

Annotation

The enhancement of Websites by personal comments is called annotation. For example, the tool Hylighter[30] allows highlighting any online text passage and to provide any note for a group of learners. Even in the very first time of the World Wide Web first research took place to combine a number of HTML-Pages to one structured online resource where users are able to search, discuss or annotate it (Dietinger & Maurer, 1998). Today, first real-life experiments are carried out where learners can share notes in real-time with other students in classrooms and therefore enhance the ongoing presentations.

Mobile Tagging

A special form of tagging is called mobile tagging, which can be described as the use of two- dimensional barcodes readable by a mobile phone (Kato & Tan, 2005). Such "mobile tags" are capable and powerful possibilities to transfer data from a physical object (such as printouts, paper) to the mobile device. The content becomes encoded to an image of small quadrates of different size and different number. This image can be printed and placed on any layer. By mobile device the image is scanned and the content is decoded by adequate software. In higher education first attempts show that there are great benefits when such

---

29 http://del.icio.us (last visit: August 2009)

30 https://www.hylighter.com/hylighter/website/index.html (last visit: August 2009)

online resources are part of the learning material. Learners can easily scan the barcode from their printouts and watch the online resource just in time on their mobile device.

Discussion Summary and Outlook

The paper states that Social Software already has a great influence and an impressive potential for the enhancement of learning within four different settings of cooperative learning settings in higher education, as we described them initially. According to the main functionalities of current Social Software – social presence, communication, collaborative development and collaborative enrichment of content – we described tools and their possibilities for cooperative learning settings. Nevertheless, there are much more tools than the named ones; the list seems to be nearly endless and steadily growing.

The use of technology in education strongly depends on the questions how we can improve the quality of education and how we can benefit from it. For example, digital collaboration with the help of Wiki systems leads to new possibilities that had not been imaginable within a paper-based learning scenario. Furthermore tagging enhances learner's content in a new meaningful way and makes the content shareable und reusable. Micro-blogging as described in previous chapters must be seen as a complete new form of communication – talking to a cloud, without knowing if anyone will read or even react to it.

As cooperative learning is very often a part of open educational practices, where learners have the possibilities to organize their own learning within their groups as active partners, changes of learning and teaching behavior is not only a matter of such new tools. Also the existing learning culture within the institution or the teaching abilities and attitudes of lectures are (amongst others) crucial aspects of teaching in higher education that has to be taken into account for a successful implementation or usage of such new tools for cooperative learning (cf. Schaffert, 2009).

Additionally, we have to bear in mind that such tools are not built especially for learning settings. It is up to the researchers and every single user to find out whether learners can benefit from it or not. Furthermore it can be expected that the number of available applications is still increasing and digital possibilities will reach new dimensions. This dramatic growth leads to the assumption that teachers and learners of tomorrow will be confronted with a huge amount of possibilities as well as more and more digital information and data. To overcome the problem of endless abundance mash-up technology will become of higher importance. Mash-ups are considered to be a concept and technology for merging content, services and applications from multiple Web sites in an integrated, coherent way (Tuchinda et al., 2008). Furthermore by studying learner habits it will be possible to recommend learners appropriate tools and content as for example first research work points out (e.g. Remashed[31]). The concept of using mash-ups combined with recommendations leads a step toward a complete personalized environment – called "Personal Learning Environment" (PLE) (Schaffert & Hilzensauer, 2008). With the aid of a PLE users should have the chance to arrange their digital environment in dependence to their personal needs. Otherwise the concept of "Cloud Computing" (use of different Web based tools) for learning purposes will not realizable.

A further very interesting aspect for digital enhanced collaborative learning will be the technology of wireless mesh networking which is already realized within the OLPC (One Laptop Per Child) project[32]. Whenever an OLPC laptop participates in a mobile ad-hoc network with other laptops it can forward packets across the network cloud. With other words computers in the cloud get automatically connected to each other and exchange date and information. Collaboration will be possible just in real time within a classroom setting, which allows working on one digital document in real time by different learners or on any other learning activity.

Beside the technology aspects, of course many questions concerning the practical experiences and real life settings are occurring. For example, even if cooperative learning is often a inspiring, funny and motivating way to learn and to teach, there are several challenges practitioners know, e.g. the "free rider"-effect, when a group member makes no contribution and let the others work, or the "sucker"-effect, when the main contributor gets more and more angry, because (s)he it the only one who works (see e.g. Renkl, Gruber & Mandl, 1996, 135ff). This means that using technology for education forces to rethink didactical approaches and the way education should be changed to improve quality.

It can be summarized that the use of Social Software has a great impact for cooperative learning and helps to create new learning and teaching scenarios. The research work of the near future targets the questions how education can take place by using different possibilities of Social Software, how technology can support the learners' and teachers' needs within cooperative learning settings, and how learners can evolve their personal learning environments by the new opportunities of networked learning.

REFERENCES

Anderson, R. J., Anderson, R., Vandegrift, T., Wolfman, S. & Yashuhara, K. (2003). Promoting Interaction in Large Classes with Computer-Mediated Feedback. In: Designing for Change in Networked Learning Environments, *Proceedings of CSCL.*, Bergen, 119-123.

Barlett-Bragg, A. (2003). Blogging to Learn. 24.01.2007, http://knowledgetree.flexiblelearning.net.au/edition04/pdf/Blogging_to_Learn.pdf (last visited: August 2009).

Bell, M. (2001). Online Role-play: anonymity, engagement and risk. In: Education Media International, Vol. *38(4)*, 251-260.

Bollen, L., Juarez, G. & Hoppe, H. U. (2006). Mobile Notes: Mobile Devices in Creative Discussions, Kaleidoscope Convergence Workshop, Amsterdam: Netherlands, http://hal.archives-ouvertes.fr/hal-00190508/en/ (last visit: August 2009).

Campione, J. C., Brown, A. L. & Jay, M. (1992). Computers in a community of learners. In: Computer-Based Learning Environments and Problem Solving. E. D. Corte, M. C. Linn, H. Mandl, & L. Verschaffel (eds.). New York: *Springer*, 163-188.

Dietinger T. & Maurer H. (1998). GENTLE – General Network Training an Learning Environment, Proc. of ED-MEDIA98 / ED-TELECOM 98, Freiburg, 274-280.

[31] http://remashed.ou.nl/ (last visit: August 2009)

[32] http://www.laptop.org/en/ (last visit: August 2009)

Dulík, T. (2009). Communication. In: K. Grodecka, F. Wild & B. Kieslinger (eds.), How to Use Social Software in Higher Education. I-Camp Project. URL: http://www.icamp.eu/wp-content/uploads/2009/01/icamp-handbook-web.pdf.

DuVall, J. B., Powell, M. R., Hodge, E. & Maureen E. (2007). Text Messaging to Improve Social Presence in Online Learning. In: EDUCAUSE, *3.*, 24-28, http://www.educause.edu/EDUCAUSE+Quarterly/EDUCAUSEQuarterlyMagazineVolum/TextMessagingtoImproveSocialPr/161829 (last visited: August 2009).

Ebner, M., Scerbakov, N. & Maurer, H. (2005). New Features for eLearning in Higher Education for Civil Engineering. In: E-Learn - World Conference on E-Learning in Corporate, Government, Healthcare & Higher Education, *Vancouver*, 635-642.

Ebner, M. & Maurer, H. (2007). Blogging in Higher Education. In: *Proceeding E-Learn*, Quebec City, Canada, 767-774.

Ebner, M. & Schiefner, M. (2008a). Microblogging - more than fun? In: Proceedings of IADIS Mobile Learning Conference 2008, Inmaculada Arnedillo Sánchez and Pedro Isaías ed., Portugal, pp. 155-159.

Ebner, M., Schiefner, M. & Nagler, W. (2008b). Has the Net-Generation Arrived at the University? oder Studierende von Heute, Digital Natives? In: Zauchner, S., Baumgartner, P., Blaschitz, E., Weissenbäck, A. (eds.), Offener Bildungsraum Hochschule, *Waxmann.*, 113-123.

Ebner, M. & Maurer, H. (2008c). Can Microblogs and Weblogs change traditional scientific writing? In: Proceedings of E-Learn 2008, Las Vegas, 768-776.

Ebner, M., Kickmeier-Rust, M. & Holzinger, A. (2008d). Utilizing Wiki-Systems in higher education classes: a chance for universal access? In: Universal Access in the Information Society, Berlin: Springer.

Ebner, M. (2009a). Introducing Live Microblogging: How Single Presentations Can Be Enhanced by the Mass. In: Journal of Research in Innovative Teaching (JRIT), 2 (1), pp. 91-100.

Ebner, M. (2009b). Interactive Lecturing by Integrating Mobile Devices and Micro-blogging in Higher Education. In: CIT - journal of computing and information technology, accepted (in print)

Ebner, M. (2009c). How Web 2.0 Enhance Knowledge Construction in Civil Engineering. In: A. Starcic, & M. Kovac. (eds.), University & Industry - Knowledge Transfer and Innovation, *WSEAS Press, Athen*, 77-101.

Garrison, D. R. (2003). Cognitive presence for effective asynchronous online learning: The role of reflective inquiry, self-direction and metacognition. In J. Bourne & J. Moore (eds.), Elements of quality online education: Practice and direction. Volume 4 in the Sloan C Series, Needham, MA: The Sloan Consortium, http://sln.suny.edu/sln/public/original.nsf/dd93a8da0b7ccce0852567b00054e2b6/755285ffb5847a4385256c3c006246ea/$FILE/Learning%20Effectiveness%20paper%20-%20Garrison.doc (last visited: August 2009)

Gräsel, C., Bruhn, J., Mandl, H. & Fischer, F. (1997). Lernen mit Computernetzen aus konstruktivistischer Perspektive. In: *Unterrichtswissenschaft*, *25*, 4-18.

Helen, S. & Wagner, C. (2006). Weblog Success: exploring the role of technology. In: *International Journal of Human Computer Studies.*, *64*, 789-798.

Holzinger, A. (2002). Multimedia Basics, Volume 2: Learning. Cognitive Fundamentals of multimedial Information Systems, Laxmi, New Delhi.

Jacksch, B., Kepp, S. J. & Womser-Hacker, C. (2008). Integration of a Wiki for Collaborative Knowledge Development in an E-Learning Context for University Teaching. In: A. Holzinger (ed.), HCI for Education and Work, *Berlin: Springer*, 77-97.

Jadin, T. & Zöserl, E. (2009). Informelles Lernen mit Web-2.0-Medien. In: bildungsforschung, Jahrgang 6, Ausgabe 1, URL: http://www.bildungsforschung.org/ Archiv/2009-01/Web2.0/

Java, A., Finin, T., Song, X. & Tseng, B. (2007). Why we Twitter: Understanding microblogging usage and communities, Paper presented at the Proceedings of the Joint 9th WEBKDD and 1st SNA-KDD Workshop.

Kato, H. & Tan, K. T. (2005). 2D barcodes for Mobile Phones, Proceedings of 2nd International Conference on Mobile Technology, *Applications and Settings*, 8.

Klamma, R., Chatti, M. A., Duval, E., Hummel, H., Hvannberg, E. T., Kravcik, M., Law, E., Naeve, A. & Scott, P. (2007). Social software for life-long learning. In: *Educational Technology & Society*, *10*, 3, 72-83.

Leuf, B. & Curringham, W. (2001) The Wiki Way. Quick Collaboration on the Web, *Addison-Wesley.*

Licklider, J. C. R., Taylor, R. & Herbert, E. (1968). The Computer as a Communication Device. *Science and Technology*, *4*, 21-41 (reprint).

McFedries, P. (2007). All a-twitter, IEE Spectrum, October, *84*.

Nagler, W. & Ebner, M. (2009). Is Your University Ready For the Ne(x)t-Generation? Proceedings of 21st ED-Media Conference (2009), 4344-4351.

Norris, C. & Soloway, E. (2004). Envisioning the Handheld-Centric Classroom, *Journal of Educational Computing Research*, *30*, 4, 281-294.

O'Reilly, T. (2005). What is Web 2.0? – Designing Patterns and Business Models for the Next Generation Software, http://oreilly.com/pub/a/oreilly/tim/news/2005/09/30/what-is-web-20.html (last visited: August 2009).

Preece, J. (2000). Online Communities: Designing Usability, Supporting Socialbility, Chichester: Wiley.

Purgarthofer, P. & Reinthaler, W. (2008). Exploring the "Massive Multiplayer E-Learning" Concept, Proceeding of 20th ED-Media Conference, AACE, 2015-2023.

Ratto, M., Shapiro, R. B., Truong, T. M. & Grisworld, W. G. (2003). The ActiveClass project: Experiments in encouraging classroom participation. In: *Computer Support for Collaborative Learning*.

Reinhardt, W., Ebner, M., Beham, G. & Costa, C. (2009). How People are Using Twitter during Conferences, In: V. Hornung-Prähauser & M. Luckmann, M. (eds.), Kreativität und Innovationskompetenz im digitalen Netz, Salzburg: *Salzburg Research*, 145-156.

Renkl, A., Mandl, H. & Gruber, H. (1996). Inert knowledge: Analyses and remedies. *In: Educational Psychologist.*, *31*, 115-121.

Safran, C., Ebner, M. & Kappe, F. (2009). Higher Education m-Learning and e-Learning Scenarios for a Geospatial Wiki, e-Learn Conference, *Vancouver, accepted*, (in print).

Schaffert, S. & Hilzensauer, W. (2008). On the way towards Personal Learning Environments: Seven crucial aspects, In: eLearningpapers, http://www.elearningpapers.eu/

Schaffert, S. (2010). Strategic Integration of Open Educational Resources in Higher Education. Objectives, Case Studies, and the Impact of Web 2.0 on Universities. In: U.

Ehlers & D. Schneckenberg (eds.), Changing Cultures in Higher Education - Moving Ahead to Future Learning., New York: Springer. (in print)

Schiefner, M. & Ebner, M. (2008). "Weblogs, more than just a toy?" or "Should I keep an e-Portfolio for my PhD study?" In: Interactive Computer Aided Learning, ICL, Villach.

Renkl, A., Gruber, H. & Mandl, H. (1996). Kooperatives problemorientiertes Lernen in der Hochschule. In: J. Lompscher & H. Mandl (eds.), Lehr- und Lernprobleme im Studium. *Bern: Huber.*, 131-147.

Salmon, G. (2002). E-Tivities. The Key To Active Online Learning, London: Kogan Page.

Slavin, R. E. (1997). Research on Cooperative Learning and Achievement: A Quarter Century of Research. *Präsentation auf dem Treffen der Deutschen Psychologischen Gesellschaft, September.*, 1997, Frankfurt.

Soloway, E. & Bielaczyc, K. (1996). Interactive Learning environments: where they've come from & where they're going, *Conference on Human Factors and Computing Systems.*, Denver (CO), 347-348.

Templeton, M. (2008). Microblogging defined, http://microblink.com/2008/11/11/microblogging-defined/ (last visited: August 2009).

Tuchinda, R., Szekely, P. & Knoblock, C. (2008). Building MahsUps by Example, ACM Proceedings of IUI 2008, Maspaloma, Spain.

Vygotsky, L. S. (1978). Mind in Society, Harvard University Press, Cambridge (MA).

Wenger, E. (2002). Cultivate Communities of Practice: A Guide to Managing Knowledge, Boston: *Harvard Business School Press.*

In: Computer-Assisted Teaching: New Developments ISBN: 978-1-60876-855-4
Editors: B. A. Morris, G. M. Ferguson, pp. 167-178

Chapter 8

COMPUTER-ASSISTED TEACHING AND NEW TECHNOLOGIES

D. Mutter [*] ***and J. Marescaux***
IRCAD-EITS – France, University Hospital of Strasbourg – France.

ABSTRACT

Many procedures in a broad spectrum of clinical specialities can be managed through a minimally invasive surgical (MIS) approach. Indeed, minimal access surgery has become increasingly accepted by the surgical community, but the surgeon is faced with the necessity of following a specific training.

Over the past two decades, advances in computer-assisted technology and telecommunications have deeply altered surgical education. The role of the traditional "Professor of Surgery" has been superseded by the quantity and quality of knowledge available not only in the media and on the Internet, but also in CD-ROMs, DVDs, stream videos, webcasts and telementoring. Nowadays, information and knowledge is readily accessible for all to find. What is more, the knowledge of highly skilled teachers is available worldwide thanks to the Internet and other multimedia technologies.

As new perspectives are opening up on the surgical horizon, such as natural orifice transluminal endoscopic surgery (NOTES), the need to apply new technologies to teaching and training as well as in daily clinical practice has become evident. This encapsulates multimedia, computer-assisted learning, videoconferencing, virtual reality and computer-assisted simulation in order to maximize surgical education. Telementoring through videoconferencing renders the acquisition of the cognitive aspects of surgical procedures easier through the observation of experts during hands-on practice. The development of computer technology allows to recreate a virtual 3D model of the patient, capable of simulating the patient's anatomy on a computer; computerized 3D images also allow virtual navigation around and into the anatomical structures, simulation of various interventions as well as automated volume calculation. In addition, a virtual reconstruction of the patient may also enhance diagnosis and enable a preoperative simulation and a preoperative training. Finally, the computerized data will allow the scheduling of automated surgical procedures. Robotics have demonstrated their ability in

[*] Corresponding author: IRCAD-EITS, 1, Place de l'Hôpital, 67091 Strasbourg Cedex, France, Phone: + 33 3 88 11 91 14, Fax: + 33 3 88 11 90 96, Email address: didier.mutter@ircad.fr

performing long-distance, remote-controlled surgical procedures. Surgeons require access to information whenever and wherever, and the Internet offers them that very chance, along with observation, training and education. Websites dedicated to surgery such as www.websurg.com incorporate all aspects of surgical training including fully illustrated and animated operative techniques, videos and experts' recommendations that, with the continuous updating of information, represent the future tool for the transfer of surgical and scientific knowledge. Using all these technologies, education may be improved through WeBSurg's virtual university as it gathers academic knowledge, education and expertise from around the world without any constraints of time and/or space. It will also facilitate the standardization of surgical treatments.

INTRODUCTION

Advances in computer-assisted technology have dramatically modified the teaching of surgery, both in theoretical and hands-on practice. Many procedures in various specialities are currently managed through a minimal access surgical approach and surgeons are faced with the necessity of adjusting their surgical training. Additionally, the conventional training modalities and working hours for residents have been reduced in many countries since it has been demonstrated that tiredness due to long shifts may lead to increased medical errors. Consequently, complementary methods of instruction are now playing an essential role in training and continuing medical education for surgeons. These new learning methods are not patient-oriented but are considered as distant or remote learning and are integrated in the surgical education curriculum. Finally, all aspects of education have been modified and the role of the traditionally well-experienced and fully fledged Professor of surgery has been eroded by the knowledge available in the medias and on the Internet, recorded on CDs, DVDs, USB keys and hard disks, but also available on video streams, webcasts and other interactive forums and chat rooms existing on the Internet. Today, information and knowledge are under no one's peer review but available to everyone seeking to find them. In addition, even the knowledge of renowned teachers is available worldwide through websites and multimedia technologies.

The advent of laparoscopic surgery in the early 90s was considered as a technological revolution. However, throughout the years, it appeared that it only represented the first step of a huge change in the surgical practice including robot-assisted surgeries and more recently, the development of Single Port Surgery and Natural Orifice Transluminal Endoscopic Surgery (NOTES). This evolution demonstrated that scars can be hidden in the natural orifices of the body, but these techniques require specific computer-assisted devices as well as dedicated training. In a near future, the combination of a computer-simulated environment promoting virtual reality and automation of surgical procedures thanks to evolved robots will offer surgeons new trends of education. These computer-generated interactive simulators will provide ideal methods for preoperative planning, repeated training of patients' data as well as sharing of surgical knowledge. A combination of these technologies will significantly alter the approach to surgical education.

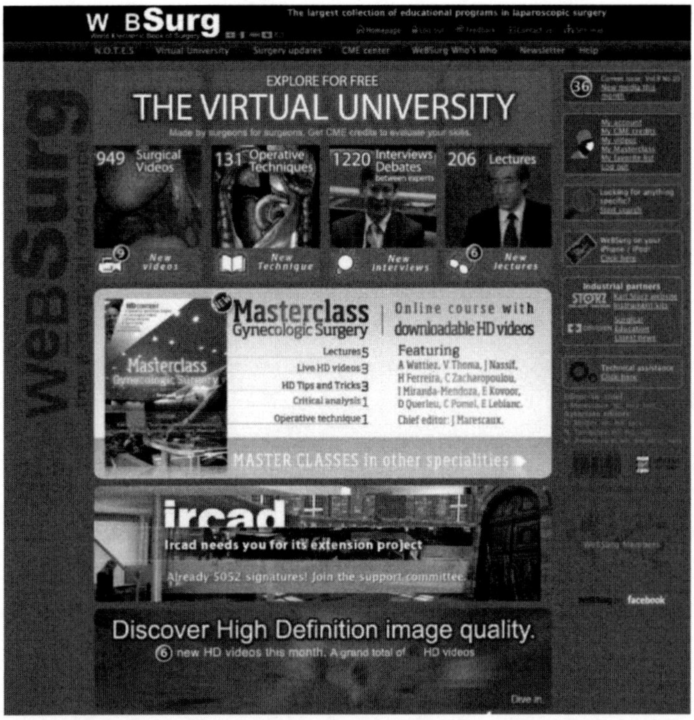

Figure 1. Virtual University www.websurg.com: home page of the website (copyright WeBSurg).

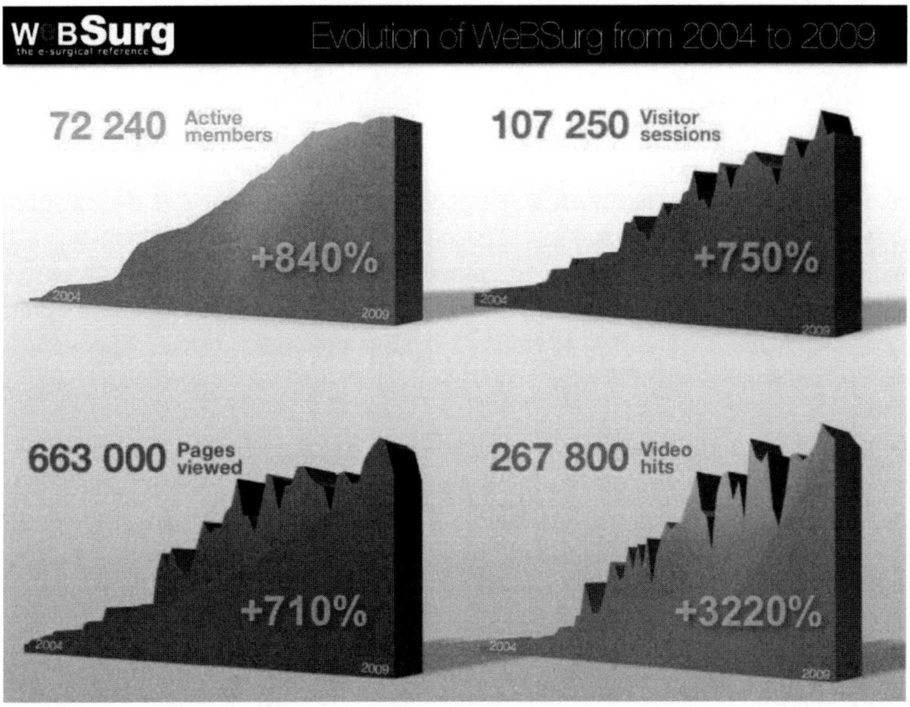

Figure 2. Statistics and evolution of WeBSurg from 2004 to 2009: Active members, visitors, pages viewed and video hits (copyright WeBSurg).

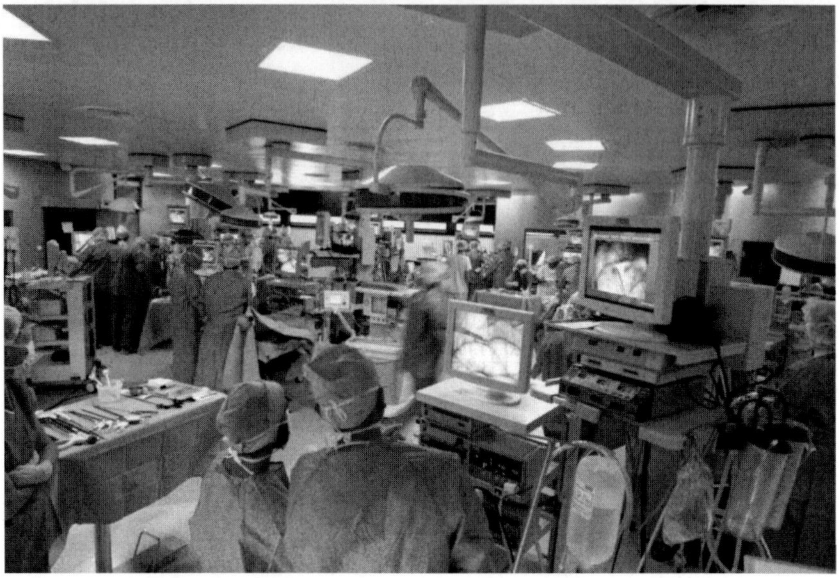

Figure 3. Experimental laboratory: introductive teaching system (picture taken by P. Eranian – IRCAD Strasbourg – France).

a. Multimedia Education

The Internet has rapidly become an essential tool for education, allowing people to stay in touch and to exchange information no matter how far apart they are. Education may benefit from the wide range of services provided by the Internet when it is delivered in the shape of high-quality data available through dedicated multimedia tools created and validated by renowned experts. The IRCAD-EITS Institute in Strasbourg, France, has developed such an Internet solution. The website, www.websurg.com (Figure 1), is one of the most complete interactive virtual Universities and is accessible throughout the world. It includes all data necessary for education in an electronic journal that is updated monthly with new videos, chapters describing minimally invasive surgical procedures, videos of surgical procedures performed and commented by a network of internationally renowned experts as well as experts' recommendations, debates and conferences on the cutting edge of surgical techniques.

At the present time, WeBSurg's virtual University provides a wide overview and a description of 131 operative techniques in minimal access surgery, 949 surgical videos, 1220 interviews and experts' debates as well as 206 lectures. Its content is organized in sequences in order to allow for an easy navigation from one core surgical step to the next one. The WeBSurg website represents an outstanding model of surgical education and Internet-based simulation. It integrates multimedia e-learning and e-training models using the latest technologies to display videos, texts and illustrations simultaneously. Membership to www.websurg.com is free, does not expire and provides its members with high quality educational data regardless of their location or finances. As a result, the number of foreign connections increased tremendously and led to the translation of the website's content from

English and French in 2001, to Spanish, Japanese and Chinese since 2008. The connections to WeBSurg for 2009 originate from Europe (32.6%), America (35.6%), Asia (25.5%), Africa (5.2%) and Oceania (1.10%). The popularity of this website is also due to its strict respect of principles and guidelines set out for health-related websites. 42 of its educational chapters are accredited to obtain CME (Continuing Medical Education) credits issued by the American Medical Association (AMA - category 1 credit™) thanks to a partnership with the McMaster University, Hamilton, Ontario, Canada. Each educational session incorporates text-based chapters on surgical interventions, videos and other multimedia tools enabling an interactive self-paced learning environment for surgeons. WeBSurg provides education in a wide range of surgical fields: general, digestive, endocrine, paediatric, gynaecologic, urological, thoracic and cardio-vascular surgery as well as interventional gastrointestinal surgery. This speciality is at the crossroads between conventional surgery and NOTES. Finally, the most recent technologies made it possible to display all the operative chapters of www.websurg.com on mobile devices such as PDAs, iPhones, iPods, and smart phones. All these factors rendered the WeBSurg website a popular and influential virtual resource that led to a 3,220 *per cent* increase in the hits of videos (from 16,000 in 2004 to 267,800 in 2009). There are 107,250 visitor sessions and 663,000 pages viewed per month. To-date, a total of 115,000 healthcare professionals are regular WeBSurg members, with a significant fidelity rate (72,240 surgeons are regular users) (Figure 2). WeBSurg demonstrated that the success of Internet-based educational resources relies on three key features: accessibility, quality of content and user-friendliness. The Internet offers a degree of interactivity that has not yet been reached by any other medium. Many surgical organizations have capitalized on it to introduce discussion forums and electronic chat rooms. Through the Internet, surgeons are able to access their favourite medical website, view videos of new surgical procedures and interact with other healthcare professionals. WeBSurg remains at the forefront through regular information updates, not only of the content but also of the technology displaying it, as well as a monthly diffusion of new videos and graphically illustrated presentations. Nowadays, the Internet and the virtual University offer adapted educational tools for surgeons in active life as well as for surgeons to-be. This type of education includes basic knowledge and advanced teaching and is recognized as part of the re-certification process of surgeons in activity [1].

b. Videoconferencing

Videoconferencing uses telecommunications of audio-video and multimedia data to gather people located at different sites for a meeting. It includes point-to-point conversation and multi-point videoconferencing involving several sites with more than one connection point. Its main interest relies in the fact that it allows the sharing of documents, computer-displayed information or any medical data of patients. Developed since the middle of the 20th century, the limited use of videoconferencing was related to its high costs, initial poor picture quality due to the lack of efficient video compression techniques and frequent connection failures. More recently, videoconferencing gained a wide acceptance in telemedicine and remote education thanks to dedicated systems allowing connections through ISDN networks. Videoconferencing evolved from highly expensive proprietary equipments, softwares and networks requirements to standard-based technologies available for everybody at a low cost. Besides, IP (Internet Protocol) based videoconferencing became the most cost-effective and reliable way to establish a connection between different places thanks to more efficient video compression technologies enabling desktop and personal computer-based videoconferencing. Basically, the core technology used in videoconferencing systems is a digital compression of audio and video through a CODEC (COde-DECode). Simultaneous videoconferencing among three or more remote points is currently possible by means of Multi-points Control Units (MCU). Data from several sources can be interconnected with high quality transmission of video and audio transmitted with high-speed Internet connectivity.

Videoconferencing provides students with the opportunity to learn by participating in a two-way communication platform. Teachers and lecturers from all over the world can participate into classes in remote or otherwise isolated places. Surgeons from various communities and backgrounds can come together to learn from each other [2] as well as to explore, communicate, analyze and share information and ideas. This original way of tele-teaching and tele-training brings about a new paradigm in medical education for teachers, learners and course organizers by breaking down the conventional constraints of time, space and learning resources. In the specific field of surgery, it is demonstrated that the acquisition of a surgical procedure's cognitive aspect is enhanced by observing an expert performing the procedure, enabling the student to gain a mental picture of the procedure as well as the ability to recognize any pitfalls. Videoconferencing, usually part of any surgical congress and course, allows many surgeons to simultaneously view high quality real-time images with an interactive sound transmission from an operating surgeon. During the transmission of the live surgical procedure, images viewed by trainees in auditoriums can potentially come from anywhere. The images displayed can be those from the local operating room, other teaching sources or even from remote sources. However, as for other types of transmissions, the quality of images represents the key point for the acceptance of such videoconferences. Therefore, the IRCAD-EITS Institute developed a local network allowing the transmission of High Definition (HD) images from the operating room to the IRCAD-EITS Institute. Images are acquired through an integrated OR1 system by HD cameras (Karl Storz®, Tüttlingen, Germany). Low compression images are transmitted from the operating room through dedicated fiber optics in an auditorium equipped with HD display screens. The incomparable quality of the images obtained allows for many hours of live videoconferencing without any visual tiredness for participants. Furthermore, the interactivity brings a lot to the experience by allowing continuous live exchanges between the operating surgeon and the participants. Consequently, live transmission is one of the most spectacular new teaching methods currently available for trainees. Through lengthy observations of live transmissions, surgeons gain surgical competencies which will in turn enable them to analyse and validate new techniques, thus improving educational communication between surgeons and experts within the surgical community. The objective of medical education is not only to achieve a high level of expertise through courses, thus compensating for the decreased number of hours in the operating room, but also to set clinical standards by assessment and confirmation of the ability to practice. The acquisition of the cognitive knowledge, competences and skills needed for surgical procedures is a fundamental aspect of surgical education. Even though first-class transmission systems including image and interactive sound, live transmissions would not have a significant impact without an efficient operator. Regardless of experience and expertise, not every surgeon can perform an educational live demonstration. The live expert acts like a movie star who should give his or her best during a live transmission. Currently, the best "live" surgeons are internationally renowned and invited to most courses and congresses. They have the ability to dissociate the surgical procedure they are performing from their comments and are able to demonstrate a procedure in a didactic way. This is especially difficult in acute situations during which the surgeon must bear in mind that sound and images are transmitted live in front of an audience. Even if they are sometimes criticized by the surgical community because of a "circus aspect", live transmissions usually encounter a huge success.

However, live transmission systems are not able to teach surgical decision-making, which represents the most important part in a surgeon's training. Indeed, the development of a surgeon's thought process goes through a faithful adherence to surgical principles, good knowledge of the medical literature and dedicated consultation with senior or more experienced colleagues.

c. Virtual Reality: Surgical Simulators and Preoperative Planning

Surgical simulation will be a major step in surgical teaching and practice in a near future [3]. Simulation will facilitate training and provide surgeons with detailed information necessary for the preoperative set-up and the surgical procedure. Developments in computer technology, robotics and Virtual Reality (VR) allow the realization of such concepts. VR is created by the interpretation of conventional two-dimensional (2D) images into digital data, leading to the formation of 3D reconstructions. These images enable the visualization of the content of structures thanks to the emergence of three new ways of perceptions: virtual immersion, virtual navigation, and virtual interaction. VR introduces a new perspective in the appreciation of anatomy as well as enhancing the interpretation of medical imaging. It has particular relevance when analyzing the relationship between a tumor and its vascular supply: therefore, it can help to optimize the balance between surgical approach and collateral damage.

Dedicated systems automatically performing the reconstruction of organs from 2D images of patient using laptop and VR software are able to delimit and recognize anatomical as well as pathological structures. Independent specialized software is used to provide 3D reconstructions. These 3D images are then viewed using surface rendering tools through an interface allowing virtual transparency of organs, virtual navigation around and into structures, simulation of various interventions and automatic calculation of volume. Vascular structures and surrounding organs can also be simultaneously reconstructed. The reconstructed organs can be evaluated and manipulated in real time. VR images obtained from patients, including those of tumors, have many applications that can be used in education.

Virtual Reality enables preoperative planning and the simulation of surgical procedures, which is of major importance in education. This initially involves automatic 3D reconstruction of the regional anatomy in order to precisely determine the anatomy of the operative field. 3D reconstruction allows to simulate several surgical approaches and strategies, which can be discussed by local or remotely located experts. Such simulations will help to optimize the position of the patients, the surgeon and port placement for the surgical procedure. These 3D models are part of computer-generated virtual models that allow for a realistic surgical simulation. Many advocate that a realistic simulation leads to better surgical training. In order to achieve this, virtual models should ideally use data from real-time situations as previously discussed. Such virtual reconstruction needs to be of high quality in order to be validated and introduced into daily clinical practice. This requires a precise and reliable system, which is able to demonstrate the surgical environment in its totality. Representation of anatomical details is essential. The accuracy and the realistic aspect of the program becomes more important according to the increasing level in surgical skills of the trainees.

Nonetheless, the most successful simulation systems are currently limited in terms of realism and appear somewhat cartoonish. On the other hand, such simulators are becoming increasingly powerful and complex and their capabilities are constantly improving. In the near future, surgical education will certainly combine theoretical courses, live transmission, video display as well as hands-on courses on live tissue and on virtual models. In fact, surgical simulation, with or without computers, is attractive as it avoids the use of patients or animals for surgical training. It ensures that trainees acquire some basic practice before treating humans.

However, the effectiveness of simulation with regards to teaching basic surgical skills (e.g., instrument and camera handling, knot-tying, and suturing) is not yet proven. For many authors, simulators measure the skills they are designed to measure [4]. Most of the studies evaluating surgical simulation (mostly computer simulation) have shown mixed results for validity in construction, prediction and reliability. The final validation is for a training simulation to bring upon positive influence on patient outcomes. In a significant review of the literature, Sutherland et al. have evaluated the effectiveness of surgical simulation compared to other methods of surgical training. Numerous simulators (MIST-VR®, Lap-Sim®, GIMentorII®, UROMentor®, VEST®), video-box trainers, physical trainers and models were compared. Most studies failed to demonstrate the superiority of the simulators to other training methods, even if they appeared better as no training at all.

Several reasons are proposed to explain this conclusion, of which sample size, disparate interventions, lack of standardization of evaluation methods, but also the fact that simulation had probably not been intensive enough. In fact, it is not possible to stay hours working on these non-realistic computer images, when long-lasting experimental procedures may be reproduced on living animals. As simulation may be just as good as other forms of surgical training, the most recreational and realistic models will be better accepted, lead to longer training, and by this probably show better results. Finally, simulators may only be interesting in validation of basic movements (rope passing, pinpointing, camera orientation) and can offer standardized and comparable data. At the present time, they cannot validate the skill of a surgeon and deliver any form of certification.

d. Telementoring

Surgical mentoring refers to a developmental relationship in which more experienced or more knowledgeable surgeons help less experienced or less knowledgeable surgeons develop specific skills and their capabilities. Mentoring involves communication, and in its organizational settings, mentoring can take many forms. In surgery, the most experienced surgeon has the objective to transmit information, thorough examples and pieces of advice to the trainee. Specific to the field of surgery is that such an exercise requires, on top of vocal exchanges, the transmission of images from a work station or from the operating room monitor to the expert through screens, computers, and in future, through hand-held mobile devices [5]. This can be carried out thanks to existing broadband, WIFI or Bluetooth technologies. If the mentor is well-equipped with the commercially available Telestrator softwares, allowing him/her to annotate or draw directly on images, the interactive discussion becomes much more interactive. This form of mentoring is probably the most accomplished demonstration of continuing education as the mentor is always available for the trainee. However, some prerequisites must be met before such a relationship can be established. It is fundamental that mentor and trainee know each other very well. Additionally, a mentor must be familiar with the working environment of the trainee in order to provide him/her with appropriate advice. The cost of this technology decreases rapidly and renders it quite affordable for trainees. However, the main problem still relies in the psychological acceptance as well as in the medico-legal problems of responsibility for the typical surgical aspect of this advice. How can the responsibility of a remote expert be evaluated if a surgeon does not follow the recommendations of an expert? At present time, surgical mentoring is mostly used in experimental studies representing one of the most effective methods of learning surgery [6]. In these studies, surgeons usually participate in surgical procedures. Such mentoring can be organized in experimental laboratories where trainees operate on animal models and are controlled, corrected and evaluated by experienced surgeons. Contrarily to computer-based training simulators in which the stepwise realization of a procedure is analyzed and evaluated, animal models allow to analyze the entire surgical procedure. This is the only way to evaluate the global performance of a procedure (such as resection and suturing of a small bowel segment) with the possibility to control its quality. For example, a bowel resection and anastomosis performed in an animal allows not only to evaluate the surgeon's gesture, but also the vascular supply, orientation, tension and viability of the anastomosis. Practical training remains essential to develop surgical skills and expertise. Indeed, traditional education based on conventional tutorial and companionship during live surgery is more and more restrictive. Consequently, hands-on courses represent an important way to acquire surgical skills. Surgeons may travel to a specialized training centre to acquire specific skill-sets pertinent to their field of surgery. This practical way of education requires the availability of high quality and local mentoring, as well as the experts' good knowledge of animal anatomy. Hands-on courses have to be organized in order to replicate the exact conditions of the operating room (Figure 3).

At the EITS experimental hands-on laboratory, surgeons work on anaesthetized 30-35 kg pigs, which are intubated and ventilated. Laparoscopic set-up includes a conventional high quality 3CCD digital camera (Karl Storz®, Tüttlingen, Germany) with conventional 0 or 30 degree angle of view telescopes. Typically, two surgeons assisted by an operative nurse and a mentor have to complete a typical surgical procedure. This procedure is chosen amongst a panel of over 35 different procedures and is part of the global educational course program. The procedure then takes place in a step by step manner, and is demonstrated by a PowerPoint slide-show presentation with drawings, pictures and short videos. These are continuously displayed on a dedicated 21-inch flat screen and are considered the guideline to the training program. The trainees successively reproduce the same procedure under the guidance of expert surgeons, who also provide vocal and practical recommendations to the trainees. Today, this practical training represents one of the most important ways to acquire surgical skills and is recognized by the ACCME Council as CME-accredited tools. They are recognized for 1 AMA category 1 credit™ per hour of training and more than 3,000 surgeons from all over the world are trained yearly in MIS techniques.

Often, during these hands-on training courses, the need for numerous experts is a serious constraint as most of the participants ask for a close and personalized expertise. Remote presence robots are proposed to increase the number of faculty participating in experimental teaching. The RP-6 robot (InTouch Health, California, USA) is a wheeled device with a mobile screen, a camera, a microphone and a driving assistance system, controlled from a computer by wireless Internet access. The instructor can thereby proctor trainees from a remote location. Such a system was successfully tested locally during an intercontinental mentoring (USA - France) [7]. These robots guarantee that remotely located experts can provide instruction to trainees and allow for interactive discussions with simultaneous image and sound transmission. Tele-presence interacts between remotely located surgeons; hospital staff and patients have already been shown to improve patient care in community hospitals.

e. Research Trends in Minimally Invasive Surgery

The continuous minimalizing of surgical scars when performing surgical procedures has led to new developments in the field of surgery, including Single Port Surgery and NOTES. These techniques represent an outstanding advance in the field of flexible endoscopy and are bridging the gap between gastroenterology and surgery. The access to the peritoneum through the transgastric route is currently evaluated in expert centres. Since 2007, many transgastric and transvaginal procedures have been performed in humans throughout the world, such as cholecystectomies and sleeve gastrectomies in obese patients [8;9]. Although in most cases, one additional laparoscopic port has been placed for safety or for technical reasons, it is anticipated that in a near future, pure NOTES surgery may become reality [10]. Training and accreditation of practitioners in this new discipline is fundamental. Existing philosophies of surgical education and practice must be altered to conquer this new paradigm. At present time, it is not clear whether interventional gastroenterologists or gastrointestinal surgeons will take the lead in NOTES. Both groups have essential skills but each needs to be introduced to the other's practice. Interventional gastroenterologists have the advanced flexible endoscopic skills, but it is the surgeons who know the terrain beyond the lumen. Therefore, new kinds of multi-specialized courses are proposed by an advanced training centre. The IRCAD-EITS Institute in Strasbourg, France, organized the first NOTES course involving interventional gastroenterologists as well as surgeons. Surprisingly, the first course was fully booked immediately after the publication of the program, although no human application had been validated. Since this first course, once basic notions were acquired, the following allowed surgeons or gastroenterologists to complete NOTES cholecystectomies in animal models, confirming the effectiveness of such an integrated educational course. Interestingly, participants frequently worked as a pair, with a surgeon and a gastroenterologist from the same hospital, showing the willingness to share their knowledge in order to perform this new surgery. The latest computer technology can aid in this process of discovery by presenting "virtual" reconstructions of the new operative landscapes to both parties. As we have seen, such technology lends itself easily to long-distance retransmissions so that the learning process does not need to be encumbered by the necessity of having the experts on the premises. In addition to acquiring procedural skills, this new field mandates the development of new instruments. Established medical technology companies such as Covidien, Karl Storz Endoscopy, J and J's Ethicon Endosurgery and Olympus are competing with newer, nimbler start-up companies in the designing, testing and production of the devices required to successfully perform these new interventions.

CONCLUSION

New technologies currently play a major role in surgical education. Internet-based education and video transmission demonstrated their ability to deliver a reliable and accredited education via its worldwide virtual University. Such a global multimedia solution at the cutting edge of technology, but also of surgical evolution, responds to the modern ethos of 'always – anywhere – anytime'. In a near future, virtual reality will offer added value to these educational tools and will be part of the training process being implemented in virtual simulators. Their effectiveness is currently limited due to their lack of realism and price, but they will probably undergo a rapid evolution once the computer limitations have been overcome. Training on cadavers and live animals cannot completely replace the limited access to operating theatres and to experts. However, due to the restricted working time and in the face of the number of surgeons looking for education in minimally invasive surgery, such training will keep a significant place in the educational process. The increasing demand for certification and validation of the acquired surgical skills by a board of professionals accounts for the multiple methods of education offered to surgeons. Each of them has the possibility to adapt his/her surgical education to the available technologies as well as to his/her practice. In many countries, the Internet has become the most important educational tool for many surgeons and remains the only way to get permanent access to new and quickly evolving technology.

We wish to express our thanks to Mr Guy Temporal for editorial assistance and valuable comments.

REFERENCES

[1] Cook, DA; Dupras, DM; Thompson, WG; Pankratz, VS. Web-based learning in residents' continuity clinics: a randomized, controlled trial. *Acad Med.*, 2005, 80, 90-7.

[2] Demartines, N; Mutter, D; Vix, M; et al. Assessment of telemedicine in surgical education and patient care. *Ann Surg.*, 2000, 231, 282-91.

[3] Szekely, G; Satava, RM. Virtual reality in medicine. Interview by Judy Jones. *BMJ.*, 1999, 319, 1305.

[4] Sutherland, LM; Middleton, PF; Anthony, A; et al. Surgical simulation: a systematic review. *Ann Surg.*, 2006, 243, 291-300.

[5] Russomano, T; Cardoso, RB; Fernandes, J; Cardoso, PG; Alves, JM; Pianta, CD; Souza, HP; Lopes, MH. Tele-surgery: a new virtual tool for medical education. *Stud Health Technol Inform.*, 2009, 150, 866-70.

[6] Takafumi, M; Tan, KY; Konishi, F; Tsujinaka, S; Mizokami, K; Sasaki, J; Kawamura, YJ. Trainee surgeons do not cause more conversions in laparoscopic colorectal surgery if they are well supervised. *World J Surg*., 2009, DOI 10.1007/s00268-009-0188-y.

[7] Sereno, S; Mutter, D; Dallemagne, B; Smith, CD; Marescaux, J. Telementoring for minimally invasive surgical training by wireless robot. Surg Innov., 2007, 14, 184-91.

[8] Dallemagne, B; Perretta, S; Allemann, P; Asakuma, M; Marescaux, J. Transgastric cholecystectomy in human beings : adapting standard instruments to new challenges. *Br J Surg*., 2009, In press.

[9] Marescaux, J; Dallemagne, B; Perretta, S; Wattiez, A; Mutter, D; Coumaros, D. Surgery without scars : report of transluminal cholecystectomy in a human being. *Arch Surg*., 2007, 142, 823-6.

[10] Mehran Anvari, JM. Natural Orifice Transluminal Endoscopic Surgery (NOTES): the dawn of a new era. *WeBSurg*., 2006.

In: Computer-Assisted Teaching: New Developments ISBN: 978-1-60876-855-4
Editors: B. A. Morris, G. M. Ferguson, pp. 179-189

Chapter 9

EXPERT COMMENTARY: E-TECHNOLOGY USE AND ABUSE IN UNIVERSITY CLASSROOMS

Rocci Luppicini
University of Ottawa, Canada.

ABSTRACT

This expert commentary addresses academic and non-academic use of electronic technology within undergraduate university classrooms. Despite the potential benefits of e-technologies, there are substantial obstacles to effective and responsible e-technology use in university classrooms that can given rise to heated debates between university administrators, instructors, and students concerning the appropriate use of e-technologies within undergraduate university classes. Based on the author's teaching and research experience working with educational technologies in Higher Education, this chapter identifies a number of serious problems and concerns revolving around current e-technology practices in university classes. This expert commentary draws on professional experience and research to provide insight into the use and abuse of e-technologies within undergraduate university classrooms. It also offers recommendations on how to leverage the educational use of e-technology within undergraduate education.

Keywords: Electronic Technology, Wireless Technology, E-Learning, Mobile Learning, Higher Education

INTRODUCTION

In order to imbue civilization with sound principles and enliven it with the spirit of the gospel, it is not enough to be illumined with the gift of faith and enkindled with the desire of forwarding a good cause. For this end it is necessary to take an active part in the various organizations and influence them from within. And since our present age is one of outstanding scientific and technical progress and excellence, one will not be able to enter these organizations and work effectively from within unless he is scientifically competent, technically capable and skilled in the practice of his own profession. — Pope John XXIII

The quote from Pope John XXIII exemplifies the push from leaders in society for greater technological integration within work and life. Technology promises but does technology deliver? Having come from an Educational Technology background, I have spent years working with instructors and students to overcome the challenge of integrating technology into classroom teaching within which instructors and students are immersed. What I have noticed over the years is that the picture of technology that is painted for us instructors working in the trenches of Higher Education is far from the actual reality of it for us or for the students we are trying to guide. This expert commentary provides a reflection on the use and abuse of e-technologies in undergraduate university classrooms. It also offers recommendations on how to leverage the educational use of e-technology within undergraduate university classes.

BACKGROUND

Laptop computers (also called notebooks) are perhaps the most popular form of electronic technology that students currently use within their university classrooms. Laptops were originally created in the early 1980s for the military as a personal computer for mobile use and small enough to sit on one's lap. It was largely a niche market until the late 1980s and early 1990s when a variety of computer companies began manufacturing increasingly sophisticated and inexpensive laptops for a broader base of users. I remember buying my first laptop from Apple's Powerbook 100 series back in 1992 when working as a research assistant in a Psychology Laboratory. It was roughly $2300 and was as heavy as a brick but was able to run a word processor and all the statistical programs I needed. It was great to have when working at the library, in coffee shops, or at my desk. As useful as it was, it lacked the wireless communication capacity, easy to use software applications, powerful broadband network technology, and the social networking opportunities available with the current World Wide Web (WWW). In addition, the price of a laptop, on average, is a fraction of what it was 10 years ago, which makes laptops much more accessible and affordable. A similar trend can be seen with other electronic technologies (I.e., PDA's, mobile phones), which have developed rapidly and grown in popularity to become widely used information and communication technologies (ICT's).

In many ways, the laptop of today is a much more complicated tool with more sophisticated computing applications and more user options than ever before. At the same time, the widespread popularity and institutionalization of laptop use within society has aligned laptops (and other mobile electronic technologies) with the assumed notion of progress and innovation. A number of private schools and some universities (I.e., Wakefield, Acadia University) have gone so far as providing laptops to each enrolled student as part of their tuition fee. Other universities (like my own) have wireless Internet access for all university members anywhere on campus and encourage instructors to embrace new technologies in classroom teaching and learning.

The rapid development of new technologies (I.e., Internet, broadband networks, new ICT's, biosensors) have transformed society and its institutions in ways that were not possible in any other period of history. Technology has helped advance many areas of society including healthcare and medical research, access to education, communications, agriculture, and transportation. At the same time, this transformation has generated new social and ethical dilemmas that did not previously exist. Technology has produced weapons of mass destruction, biological warfare, environmental contamination, and cybercrime. This double-edged sword has given rise to what I refer to as the *paradox of technology* (Luppicini, 2008). On one hand, individuals and institutions depend on technological innovation to improve life and society; on the other hand, technology can have many anti-human consequences that individuals need to oppose in order to protect themselves.

Within the context of Higher Education, concerns about the use of technology to leverage educational innovation and progress must be juxtaposed with concerns about possible negative influences of technology on the educational context and the user. To this end, the following discussion reflects on key challenges I have encountered over the years in efforts to integrate electronic technologies into university classroom contexts. More specifically, it draws on student-instructor convervations and classroom observations to offer insight into challenges and opportunities arising in wireless classroom contexts.

REFLECTIONS FROM THE ACADEMIC TRENCHES

I am a professor at a large 30,000+ urban public university in Canada and have been at this current institution since 2005. At my university, some professors are using a number of technologies in their classroom teaching (I.e., interactive whiteboards, clicker student response systems, PDA's, blogs, podcasts, digital photography, video conferencing) to help nurture stimulating learning environments for their students. Other professors are using technologies as a crutch to reduce their teaching efforts and workload, as is often the case when lectures consist of literally reading overhead projecter or PowerPoint presentation slides. And then there are professors who refuse to use new technologies in their classroom teaching because they believe it interferes with their teaching. What I have realized as both a student and instructor is that technology cannot replace good teaching and what instructors excel at: providing rich learning environments, motivating students, scaffolding student learning, explaining and demonstrating concepts, and engaging students in learning. Technologies, however, do offer more options to augment good teaching, which is at the core of effective instruction. This is particularly salient in the era of WiFi where many universities offer wireless connections to students within classrooms and on campus. The question is: How are electronic technologies being used (or abused) by students and instructors within university classrooms. This section deals with the paradoxical situation that many students and instructors face within the university classroom of the future. As will be seen, the technology is a double-edged sword with both positive and negative sides that must be appreciated before one can decide its overall contribution to university teaching and learning (if at all).

Meet the Net Generation

A colleague of mine who just began her university teaching career had a female student approach her at the end of class to complain about offensive material seen on the laptop of a male student sitting directly in front of her in a large undergraduate class. The female student was quite bothered and demanded that something be done. The next class, my colleague walked back through the auditorium to see what the problem was. On the screen of one of the students was a large image of a man holding onto the hair of a woman while ejaculating into the woman's face. Underneath the graphic scene was written in bold, "If you won't swallow—you can take it in the eye!" When the class was dismissed my colleague asked this male student to stay behind for a moment to discuss appropriate classroom conduct. When confronted with this complaint the student protested, "I paid my tuition and it is my right to use my laptop in class however I want. It is not my problem if other students are uptight. I am not complaining about them!" These are the problems of our times but what are the solutions, if any?

In spite of the remarkable benefits of electronic technologies, research on the role of technology in organizational settings has found that individuals do not always use electronic technologies effectively and responsibly. Misuses of electronic technology within organizational settings (I.e., cyberloafing, network attacks, information theft, unauthorized transmission of confidential information, cyberbullying, exchanging online pornography) are a serious problem in contemporary organizational life. Within Higher Education, the misuse and abuse of electronic technologies within colleges and universities have led to classroom distraction (Bugeja, 2007), low level learning when multitasking (Hembrooke & Gay, 2003), classroom disruptions, and forced instructor bans on electronic technologies in the classroom (Cole, 2008).

Over the last few years I have conducted in-class research studies to uncover student perspectives on electronic technology use and abuse in the classroom. Through this work and many in-class discussions with students, I have found that overall, students are aware that e-technology abuse within university classes takes away from their learning but admit to doing it anyway. Many students have commented that despite their best intentions, the temptation to misuse technology in the classroom is tough. One student confided, "I always start out taking notes, but usually end up surfing the web. Sometimes I can wait until class is done, but it's right there and it's so easy and quick that sometimes I do it". My own classroom observations are validated by research focusing on the distracting force that technology can have on university classroom learning. As stated by Bird, Cain, and Jones (2008):

> "In addition to the more traditional types of distraction such as crossword puzzles, mobile computing devices offer online shopping, instant messaging, Facebook, online fantasy sports games, and web browsing. Aside from the general annoyance this may create for instructors, computer-aided distraction has the potential to negatively impact learning" (p. 2).

The paradox of technology in Higher Education teaching contexts is evident. On the one hand, technology is a powerful tool that can leverage teaching and learning. On the other hand, technology is a distracting force in classrooms and competes with instructors for student attention. As I see it, this paradox is not possible to get away from because it permeates the lives of our students who are the net generation and who have grown up in a culture of technology (Tapscott, 2008).

The Paradox of Educating the Net Generation

Perhaps the biggest problem with technology (and ourselves) in our contemporary society is learning to live with contradiction. Chickering and Ehrmann's (1996) Implementing the Seven Principles: Technology as Lever is one of a number of popular sets of guidelines adapted by teaching and learning support units at numerous North American colleges and universities. It has been used as a tool to promote ways to use electronic technologies within undergraduate classes. I am using these principles as an illustration of one side of the double-edged sword of technology applied to teaching in Higher Education. The seven principles are as follows:

1. Good practice encourages contacts between students and faculty

Communication technologies that increase access to faculty members help them share useful resources, and provide for joint problem solving and shared learning can usefully augment face-to-face contact in and outside of class meetings.

2. Good practice develops reciprocity and cooperation among students

The increased opportunities for interaction with faculty noted above apply equally to communication with fellow students. Study groups, collaborative learning, group problem solving, and discussion of assignments can all be dramatically strengthened through communication tools that facilitate such activity.

3. Good practice uses active learning techniques.

The range of technologies that encourage active learning is staggering. Many fall into one of three categories: tools and resources for learning by doing, time-delayed exchange, and real-time conversation.

4. Good practice gives prompt feedback

The ways in which new technologies can provide feedback are many — sometimes obvious, sometimes more subtle. We already have talked about the use of email for supporting person-to-person feedback, for example, and the feedback inherent in simulations. Computers also have a growing role in recording and analyzing personal and professional performances. Teachers can use technology to provide critical observations for an apprentice; for example, video to help a novice teacher, actor, or athlete critique his or her own performance.

5. Good practice emphasizes time on task

Technology also can increase time on task by making studying more efficient. Teaching strategies that help students learn at home or work can save hours otherwise spent commuting to and from campus, finding parking places, and so on.

6. Good Practice Communicates High Expectations

New technologies can communicate high expectations explicitly and efficiently. Significant real-life problems, conflicting perspectives, or paradoxical data sets can set powerful learning challenges that drive students to not only acquire information but sharpen their cognitive skills of analysis, synthesis, application, and evaluation.

7. Good Practice Respects Diverse Talents and Ways of Learning

Technological resources can ask for different methods of learning through powerful visuals and well-organized print; through direct, vicarious, and virtual experiences; and through tasks requiring analysis, synthesis, and evaluation, with applications to real-life situations. They can encourage self-reflection and self-evaluation. They can drive collaboration and group problem solving. Technologies can help students learn in ways they find most effective and broaden their repertoires for learning (p. 4-5).

Juxtaposing the seven principles with real life university teaching situations reveals paradoxical findings with respect to the abovementioned principles. First, good practice does not encourage contacts between students and faculty. It is often the case that increased access to faculty leads to high-frequency low-quality communications and increased miscommunication. During the first few weeks of class each terms, instructors can spend many unwanted hours sifting through repetitive email questions, many of which concern student service matters and other general questions that do not pertain to academic work. Other times unwanted spam is sent by students (often unknowingly) when their email list is compromised by a third party entity (I.e., requests from social network sites to join). And although the frequency of student-instructor communication may increase, it is often low-quality communication that requires more effort and patience from instructors. Too often multiple messages are posted by the same student with each message sent to clarify some error in the previous message. The formal quality of message content itself is all too often lacking with instructors referred to as "Dude" or "Hey You". It is doubtful that such a use of technology can foster student instructor relationships that will enhance students' intellectual commitment and encourage them to think about their academic values and plans. In order for the practice to be a "good" practice, it must follow organizational guidelines established by the professor and follow communication protocols governing electronic communications. The reformulated principle is thus: *Good technological practices that encourage contacts between students and faculty are conditional upon the presence of appropriate organizational strategies and rules of netiquette.*

Second, in reflecting on my own teaching experiences and discussions with colleagues, I have found that working with others in online collaboration often decreases student involvement in learning. Some colleagues of mine have complained about online 'lurkers' who do not engage and simply copy the final product worked on by others. When this instructor assigned points for online participation, the lurkers switched strategies by making irrelevant or superficial comment to receive participation points. While it is true that sharing ideas and providing feedback can deepen student understanding, the increased opportunities for student interaction using communication technology can dramatically weaken student motivation and learning. This is partly due to removal of face-to-face contextual cues and a diminished sense of social affiliation and responsibility that often plagues online group work (see Luppicini, 2006). The technological mediation of student interactions can lead to psychological distancing effects and competition among learners. In other works, collaboration cannot be assumed. Rather, it must first be cultivated and situated within participatory networks and learning communities (Lave & Wenger,1991; Luppicini, 2007;

Wenger, 1990). *Good technological practices that foster reciprocity and cooperation among students are conditional upon the establishment of a learning community or participatory social network within an explicit learning context.*

Third, despite the range of sophisticated electronic technologies and software available to institutions of Higher Education, their application often fails to encourage active learning. For instance, our university recently began introducing clicker response systems within large university classes to encourage greater student involvement in classroom discussion. Although some of my colleagues have used it effectively, other admit to using only to record attendance and noting else. In other words, it is not what the technology is capable of doing but rather, what is being done with it. From the student perspective, the decision to treat the classroom experience as a spectator's sport or something more depends not so much on the technology but how well the instructor can use it to get students to actively participate in class discussions in meaningful ways that connect to what they are learning, how what they are learning relates to past experiences, and how this might apply it to their daily lives. With or without the assistance of electronic technologies, the instructional aims are the same-- to actively engage students in their learning. *Good technological practices that foster active learning among students are conditional upon effective instructor strategies to actively engage students in their learning.*

Forth, electronic technologies often fail in providing prompt feedback that is relevant to the learner and the learning context. Rather than offering student opportunities to reflect on what they have learned and what they still need to know, technology mediated feedback may be unstructured, unorganized and low in quality. Although, electronic technologies can be used to mediate feedback in many ways between multiple users, this does not mean that the feedback is useful. It all depends on how the instructor regulates the use of feedback mechanisms that the technology offers and how students respond to this. In one of the first university undergraduate classes I taught, I integrated online peer feedback into an undergraduate course in Communications using a peer feedback template for individual class presentations. As a precaution I decided to screen the feedback before forwarding it on to the student being critiqued. In this all female class of new undergraduate students, inappropriate comments like, "You are too fat to wear tight clothes while presenting" and "Your poor sense of fashion detracts from your presentation" were scattered throughout the feedback. The lack of prior face-to-face group work and relation building classroom exercises allowed the distancing effects of technological mediation to influence the feedback. Whether using ICT's for supporting person-to-person feedback, or using multimedia to record and analyze classroom performance, *technological practice promoting effective student-student and instructor-student feedback is conditional upon how instructors regulate the use of available feedback mechanisms.*

Fifth, new technologies can dramatically decrease time on task for students and faculty members. Instead of using technology to work on course projects, my students have often discussed technology as their major distracter from completing class assignments. Students taking a break to "surf the net" while writing a paper often turn in papers late. And during class time, it is not abnormal to answer the same question multiple times to reach students who were too busy texting on their mobiles to pay attention to the lecture content. In this way, technology can decrease time on task by introducing new distractions for students to cope with. When Facebook was first launched, multiple students and colleagues alike admitted that they lost entire workdays finding out the whereabouts of childhood friends, old schoolmates, and extended family. *Good technological practices promoting time on task are conditional upon teaching strategies encouraging time on task technology use and discouraging time off task technology use.*

Sixth, new technologies can fail to communicate teaching and learning expectations. Overreliance on Internet resources can limit time available to analyze real-life situations and alternative perspectives from sources not found on the Internet. In discussing contested marks with students with poorly researched papers, I am continually surprised at how many students admit to only using readily available Internet sources from the Google search engine for research papers without delving into university libraries or database holdings. In this sense, technology can actually lower expectations on information acquisition and dull cognitive skills of analysis. This "good enough" approach to research is creating major learning challenges for students who are often relaying on whatever is found in the Internet's digital garbage dump and not learning how to acquire information through other available means. *Good technological practices that fulfill high teaching and learning expectations are conditional upon the understanding of learning aims and objectives driving teaching and learning expectations.*

Seventh, technological practices can compromise respect for diverse talents and ways of learning. Technological reliance can overemphasize some methods of learning (E.g., multimedia tools, asynchronous communications, virtual experiences, simulations, etc.) at the expense of other learning methods requiring the analysis and evaluation of real-life situations in real-time. The flexibility and convenience of ICT's can discourage face-to-face collaboration and group problem solving, which in turn, can limit student learning repertoires. I recently had a student in her fourth year of an honors undergraduate program that failed her core research course and wanted to turn in extra work to receive a passing grade. When asked about her literature review she stated," I do not know where I went wrong. My Google search turned up 20 sources to use in my literature review!" This student admitted proudly that she had never logged on to the library database or set foot in the library in her four years attending the university. The fact that she passed all of her course up until this research course is troubling. Even more troubling is the narrowing of learning options and resources that this student and others are relying on. *Good technological practices that promote respect for learning diversity is conditional upon effective instructional strategies that instill respect for learning diversity.*

RECOMMENDATIONS ON THE WAGING BATTLE OVER TECHNOLOGY IN HIGHER EDUCATION

It is not enough to have all the bells and whistles that technology offers at the expense of sound pedagogy. Based on the reflective analysis offered in this paper, the following seven principles are instructor-centered strategies for technology integration into classroom settings:

1. Develop good organizational strategies and rules of netiquette when introducing technological practices to leverage contact between students and faculty.
2. Cultivate a learning community within an explicit learning context when introducing technological practices to promote reciprocity and cooperation among students.

3. Develop effective strategies to foster active learning among students when introducing technological practices to leverage active learning in the classroom context.
4. Develop effective strategies to regulate the use of available feedback mechanisms when introducing technological practices to facilitate student-student and instructor-student feedback.
5. Develop teaching strategies to encourage time on task technology use and discouraging time off task technology use when introducing technological practices to leverage learning efficiency and time on task.
6. Develop strategies to leverage student understanding of learning aims and objectives when introducing technological practices to satisfy high teaching and learning expectations.
7. Develop effective instructional strategies that instill respect for learning diversity when introducing technological practices to promote the same values.

Rather than focusing on the adaption of pedagogy to technology integration demands, these principles highlight the adaption of technology integration efforts to instructor pedagogy. The intention is to highlight the priority of professional teaching practice as drivers of technology integration into teaching and learning. This, in turn, may also help instructors to discover new teaching strategies and methods to win back the attention (and respect) of their students who have grown up as the net generation within a distracting culture of technology.

DISCUSSION: TECHNOLOGY IS NOT ENOUGH

Technology is a double-edged sword with the capacity to leverage or lower the quality of university education. This paper has argued that when faced with course requirements that use technologies in various ways, instructors must place organizational considerations and teaching strategies in the forefront of any technology integration effort. The principles offered in the discussion are intended place technology in the service of instructors rather than the other way around. Faculty members who work with technology should make sure it is being used in the service sound pedagogy to promote student learning. Institutions of Higher Education that wish to promote greater technology integration within university classrooms have a dual challenge to: (1) address technology support needs of their constituency, and (2) to address professional development needs of instructors faced with the challenge of trying top effectively use technology into their university classrooms to augment teaching and learning practice. Although this paper highlighted the paradoxical nature of technology applied to Higher Education, more work is needed to provide a detailed inquiry into the underlying conditions and factors driving paradoxical outcomes of technological integration within various contexts of Higher Educations.

In terms of competing with technology for student attention within university classrooms, the instructor is faced with an ongoing challenge that is not easy to overcome. Regardless of the instructor response to technology abuse in the classroom (E.g., restricted use policies, classroom bans on e-technology, no action, etc.) the distracting force of electronic technology cannot be eliminated outside the university. It will resurface wherever electronic technology is present, be it in the university library, coffee shop, or at home. As a university instructor, I have found that the best way to deal with this situation is not through restrictive bans and anti-technology use policies. This only reduces class attendance and promotes secrecy and deception. It is impossible to patrol a large undergraduate class of 400+ students effectively to find out which students may be text messaging on their mobiles. I might add, it is not our job to act as a police or babysitter to young adults. The best way around this in my opinion is to create a stimulating class environment that students want to attend. It is a challenge to university instructors but one that many of us are taking each day.

ADDITIONAL READING SECTION

Adams, D. (2006). Wireless laptops in the classroom (and the sesame street syndrome). Association for Computing Machinery.Communications of the ACM, 49(9), 25.

Campbell, Andrea B., and Pargas, Roy P. (2003). Laptops in the classroom. *Technical Symposium on Computer Science Education archive*. Pp. 98-102. New York, N.Y., USA: ACM.

Campbell, S. (. (2006). Perceptions of mobile phones in college classrooms: Ringing, cheating, and classroom policies. Communication Education, 55(3), 280-294.

Carnevale, D. & Young, J. R. (2006). The challenges and benefits of requiring students to buy laptops. The Chronicle of Higher Education, 52(38), A39(1)-40.

Gilroy, M. (2004). Invasion of the classroom cell phones. Education Digest: Essential Readings Condensed for Quick Review, 69(6), 56-60.

Lavoie, J. A. A. & Pychyl, T. A. (2001). Cyberslacking and the procrastination superhighway: A web-based survey of online procrastination, attitudes, and emotion. Social Science Computer Review, 19(4), 431-444.

Nilson, L.B., Weaver, W. E. (2005). Laptops in Class: What Are They Good For? What Can You Do with Them? *New Directions for Teaching & Learning, Issue 101*, 3-13

Poldrack, R. (2006). iPods and Learning: Not Always a Good Mix. *American society for training and development, 60*, 14-14.

REFERENCES

Bird, E. R., Cain, J. & Jones, M. (2008). Mobile Computing Initiatives Within Pharmacy Education. *American Journal of Pharmaceutical Education, 72*, 1-7

Bugeja, M. (2007) *Distractions in a wireless classroom.* The chronicle of Higher Education. Retrieved November 15th from http://fp.arizona.edu/geog695c/PDFs/wireless.classroom.pdf

Chickering, A. & Ehrmann, S. (1996). Implementing the Seven Principles: Technology as Lever. *AAHE Bulletin,* October, 3-6.

Cole, David. (2008, October 23). Why I ban Laptops in my Classroom. *Encyclopedia Britannica Blog.* Retrieved November 20, 2008 from http://www.britannica.com/blogs/2008/10/why-i-ban-laptops-in-my-classroom/

Hembrooke, H. & Gay, G. (2003). The laptop and the lecture: The effects of multitasking in learning environments. *Journal of Computer Higher Education., 15*, 46-64.

Lave, J. & Wenger, E. (1991). *Situated learning: Legitimate peripheral participation.* Cambridge University Press.

Luppicini, R. (2006). Review of Computer Mediated Communication Research For Education. *Instructional Science, 35*, 141-185.

Luppicini, R. (2008). Introducing Technoethics. In R. Luppicini & R. Adell (Ed.), *Handbook of Research on Technoethics: Volume I.* Hershey: Idea Group Publishing.

Luppicini, R. (Ed.). (2007). *Online Learning Communities.* Charlotte, NC: Information Age Publishing.

Tapscott, D. (2008). Grown Up Digital: How the Net Generation is Changing Your World. Toronto: McGraw-Hill.

Wenger, E. (1999). *Communities of practice: Learning, meaning, and identity.* New York: Cambridge university press.

In: Computer-Assisted Teaching: New Developments ISBN: 978-1-60876-855-4
Editors: B. A. Morris, G. M. Ferguson, pp. 191-200

Chapter 10

PARTICIPANTS IN THE COMMUNITY OF KNOWLEDGE: EXPERIENCES WITH A WIKIPEDIA WRITING ASSIGNMENT FOR STUDENTS

Normann Witzleb
Faculty of Law, Monash University, Australia.

ABSTRACT

Improving students' computer literacy and instilling a critical approach to internet resources are important educational aims for today's generation of students. This Commentary will examine how a writing exercise in the style of a Wikipedia article can be used to facilitate these skills. Students in an elective unit in Comparative Law were asked to write, and review, an article on an issue, concept or scholar in Comparative Legal Studies. The Commentary will describe the rationale for adopting this form of assessment, how it was integrated into the teaching and assessment structure of the unit, and how students responded to the exercise. It will critically evaluate this novel form of assessment. The aim of the Commentary is to provide some practical guidance on the potential benefits of this form of assessment and when it might be usefully employed.

INTRODUCTION

Wikipedia is a multilingual web-based free encyclopedia which is written and edited by countless contributors from around the world. Its use by students continues to divide academics. In principle, articles in Wikipedia are open to be changed by any user with access to the Internet, simply by clicking an edit button on each website. This 'open content' design differs radically from conventional sources of academic learning, such as textbooks and articles in scientific journals, which usually undergo rigorous editorial or peer-review processes before publication. The way in which Wikipedia controls the quality of its content is not through commissioning experts to write or referee, but through a watchful army of interested persons from all walks of life and all level of ability and knowledge. Despite occasional (often only temporary) lapses in accuracy, the overall quality of much content in Wikipedia, in particular rated articles, is increasingly acknowledged (Wedemeyer, 2008). It is certainly now more comprehensive and, often, also more current than many traditional encyclopedias.

Some academics fear that the existence of Wikipedia undermines academic standards, leads to poor scholarship and uncritical use of its content (Geiger, 2008). This criticism is, of course, not limited to students' use of Wikipedia alone. It concerns the way in which students employ most information provided in electronic format, be it on the internet or on more subject-specific databases. However, there is no doubting that, since its creation in 2001, Wikipedia has grown rapidly in its size and user traffic. Wikipedia now has nearly 3 million articles[1] and has become the largest reference website on the Internet.[2] Recent studies confirm that Wikipedia holds particular appeal for younger users. In a recent German internet usage survey, 94% of the teenagers between 14 and 19 years indicated that they used Wikipedia at least sometimes, compared with 65% of all users and 35% of users of 60 years of age (Busemann & Gscheidle, 2009). The ubiquity of Wikipedia has led teachers to explore ways in which students can be educated in using it safely and responsibly (Jenkins, 2007). One way of doing so is transforming students from uncritical consumers of Wikipedia to creative contributors. This is in line with the basic rationale of 'wikis', a term used for collaborative websites that allow for easy creation and editing of content by a group of users.[3] Encouraging students to become 'Wikipedians' would also help to bridge divide between active and passive users of Wikipedia, with only 4% of users actively contributing to the development of the site (Busemann & Gscheidle, 2009).

This chapter will describe the use of Wikipedia in the assessment of an elective subject in an undergraduate law degree at an Australian university. Setting such exercises is no longer pioneering.[4] Wikipedia's 'School and University projects' website[5] currently lists about 75 past and over 30 current projects worldwide in which educators, generally at tertiary level and most often in North America, have used Wikipedia as a teaching and assessment tool. To encourage and to provide guidance in the preparation of such initiatives, the Wikipedia Foundation now provides resources to assist teachers in incorporating Wikipedia writing assignments into their classroom practice.

Rationale and Design of the Project

The exercise was prompted by a number of considerations. First, as experts in a field, we often underestimate how difficult it is to find a foothold, and a sense of direction, in new terrain. Introducing others to the fundamentals of an area is a challenge. It holds valuable

---

[1] See Wikipedia: Size of WIkipedia, http://en.wikipedia.org/wiki/Wikipedia:Size_of_Wikipedia (December 4, 2009).

[2] Wikipedia is among the ten most popular sites on the Internet and the most widely used reference web site; see Alexa Top 500 Global Sites, http://www.alexa.com/topsites (December 4, 2009).

[3] The name Wikipedia is a portmanteau of the words *wiki* (from the Hawaiian wiki-wiki, meaning quick) and *encyclopedia*.

[4] At the time of setting, the exercise (semester 2, 2007) appears to have been among the first to use wikipedia for assessment in a law course.

[5] See Schools and Universities Project, http://en.wikipedia.org/wiki/Schools_and_universities_project (December 4, 2009).

lessons also for students to provide an account of the state of knowledge in a particular area and it prepares a strong basis from which to conduct further research and analysis. Second, the coverage of standard textbooks in Comparative Law did not coincide with the syllabus of the unit and they no longer fully reflected current knowledge. Introductory material in this specialized field of study was thus quite difficult to come by. Students researching and producing it themselves had therefore immediate and tangible benefit for them. The exercise went thus beyond asking students somewhat aridly to produce a further iteration of readily available material. Third, requiring students to write on these topics for an online medium also contributed to providing further information on these topics to a wide audience. It is peculiar that the internet has quickly developed into one of the indispensable sources of information on foreign law (Witzleb, Martiny, Thoelke, & Frericks, 1999), yet the availability of electronic information on comparative law as a subject, its key concepts and major scholars has not kept pace with this revolution. This assessment exercise was intended, in a small way, to address this imbalance and give comparative law scholarship some more prominence in mainstream Internet fora. Fourth, the exercise was designed to give students an authentic task that educated them in the self-reflective use of secondary sources and, at the same time, enhanced students' computer literacy and research skills at an advanced stage of their degrees. Finally, the exercise was also influenced by teacher's own recent experience of contributing to a conventional encyclopedia,[6] which had caused me to reflect on the demands, and sense of achievement, that writing for a reference works can bring about.

The assessment was divided into two parts. The first component of the exercise consisted in students drafting a Wikipedia article. Students could choose a topic from a list with suggestions or write on an alternative topic after submitting it for approval. In the second half of the exercise, students were asked to review an article written by one their peers. In other words, the exercise was designed to require students to work on two distinct topics. Each student wrote an original article and revised the work of another student on a different topic.

Instructions Given to Students

Students were allocated a topic from a list of potential articles. At the time of creating the list, most of these articles were either non-existent or existed only in embryonic form ('stub'). The list of suggested topics was diverse. Students had the opportunity of writing on a important concept, movement or an influential scholar or work in Comparative Law. Fundamental concepts included: Common Core Methodology, Micro-/Macro Comparison, Mixed Legal Systems, Legal Transplants, Legal Families, Legal Culture, Globalisation of Law, Unification of Law, Harmonisation of Law, Transnational Law, European Contract Law, European Tort Law, European Family Law and European Civil Code. Students who were assigned to write a biographical articles need to research into the life and work of eminent comparative law scholars, such as Ernst Rabel, René David, Rudolph Schlesinger, Max Rheinstein, F.H. Lawson, and H.C. Gutteridge. A final set of topics gave students the opportunity to prepare an account of influential foreign legal codes, such as the Dutch Civil

[6] In *Brockhaus Enzyklopädie*, 21st ed. (2005-6), with 30 volumes the most comprehensive German-language encyclopedia.

Code, the Italian Civil Code or the Spanish Civil Code. The fact that none of these topics had an entry that went beyond rudimentary content ensured that all students had a similar starting point. All articles had to be designed and researched from the ground up, and no student could (or had to) consider preparatory work done by others. Students who were nonetheless unhappy with their allocated keyword, had the option to suggest writing on a different topic and, after receiving approval, research the area of their choice.

Each topic was assigned twice but students had to work independently. Students were referred to the information available on how to write and edit for Wikipedia, which enabled also those unfamiliar with Wikipedia to understand the task involved. Wikipedia itself contains a comprehensive range of clear and continuously updated guidelines on article writing. These guidelines address writing 'Your first article',[7] they include a 'Guide to writing better articles'[8] and even explain how to write 'The perfect article'.[9] Students were instructed to comply with the Wikipedia standards and policies.[10] It was suggested to students that they write for submission to Wikipedia but they were not obliged actually to upload the article. Assessment criteria for this component of the exercise were the content of the article as well as its style, in particular its compatibility with the Wikipedia policies and guidelines. This assessment item counted for 15% of the overall mark in the unit.[11]

The second component of the exercise consisted in writing a review of a draft article. It is characteristic of Wikipedia that it is a collaborative effort, and this assessment item aimed to mimic this feature of Wikipedia. Students were asked to assess the quality of a draft Wikipedia article written by one of their peers and, where appropriate, to suggest amendments. Where a student suggested revisions he or she was expected to explain why they would improve the article. Two criteria for assessment of the review article were stated: Firstly, the degree to which the content and style of revisions were in line with the Wikipedia policies and guidelines, and secondly, the extent to which the revisions and comments served as helpful guidance in the assessment of the draft articles. This assessment item counted for 5% of the overall mark in the unit.

Assessment of Students' Work

The assignments were reviewed and marked, as far as possible, anonymously. Articles were returned with teacher comments as well as the marked review. Reviews were returned with teacher comments as well as the marked articles. Students were encouraged to provide candid feedback but at the same time reassured that their assessment of their peers' work was by no means determinative. Having the reviews was of great assistance in the marking process. Even though the assessment instructions stated that the comments received from

[7] Wikipedia: Your first article, http://en.wikipedia.org/wiki/Wikipedia:Your_first_article (December 4, 2009).

[8] Wikipedia: The perfect article, http://en.wikipedia.org/wiki/Wikipedia:The_perfect_article (December 4, 2009).

[9] Wikipedia, Guide to Writing Better Articles, http://en.wikipedia.org/wiki/Wikipedia:Guide_to_writing_better_articles (December 4, 2009).

[10] Students were referred to the policies of Wikipedia at http://en.wikipedia.org/wiki/Wikipedia:List_of_policies (December 4, 2009) and other relevant wikipedia websites.

[11] Other assessment items were the review of the draft article (5%), see below, a critical review of a journal article (15%), a research assignment (35%) and a take home exam (30%).

fellow students were not binding, the idea of giving and receiving peer feedback needed some getting used to. The following student comment may well be representative:

> Reviewing another student's article was a novel experience. I think some students were cynical because they thought their mark would be a reflection of the quality of the other student's work. But it is rare you get to read what another student has written in its entirety, and it gives perspective to your own work. The process of critically evaluating another piece of work is a process that can easily be translated to reviewing your own work, and so should improve your own writing.

The articles for review and revision were allocated by the teacher, thereby ensuring that students gained wider exposure to fundamental concepts of the discipline area. The success of this strategy is evidenced by the following student feedback:

> Reviewing another student's article exposed us to another topic of Comparative Law. In that sense it was a great way to get us thinking about Comparative Law concepts - since obviously an introductory unit cannot hope to cover everything.

The standard of the submitted articles and reviews was, overall, very pleasing. A fair number of students had evidently developed great enthusiasm for the task, and their respective topics, and delivered excellent work. On the other hand, some other students' work had weaknesses that any form of student writing may suffer from. Such generic criticism related, for example, to the quality of research, the clarity of writing and lack of care taking with editing and referencing. Sometimes, factual statements were inaccurate, incomplete or out-of-date. Where students relied too heavily on a limited number of sources, their article tended to be patchy in its coverage or present a distorted view of the subject matter. Other common complaints concerned the clarity of prose or structure. Even though these deficiencies can often also tarnish conventional research essays, they were more concerning here. To be taken seriously, an encyclopedia needs to strive for unassailable quality and meticulous exposition. Accuracy, comprehensiveness, clarity and attention to detail were key but not all students had approached the task with that ambition, or succeeded in their efforts.

Most instructive for the current context is how students handled the conventions specific to Wikipedia. Some students evidently had difficulty coming to grips with the specific demands of writing in a novel format, and for an unfamiliar purpose. It became evident that not all students read the instructions or followed them. Broadly speaking, Wikipedia adopts many of the conventions of a traditional encyclopedia. Articles should begin with a lead section, which should contain the shortest possible definition or description of the subject-matter. The lead section should be able to stand alone and give a concise overview of the article. Often, a summary style is preferable, which means that important information be dealt with first, then a summary be given before the topic is explored in further detail. Some students also had difficulty finding the right tone for the article. The articles should be pitched at the educated layperson and adopt a neutral point of view. Writing should be concise and self-contained. Some of the guidelines for proper style contain valuable lessons also for other forms of legal writing.[12]

---

[12] For example, wikipedia cautions against the use of "Weasel Words" (such as "it is widely regarded as" or "some have claimed"), through which an author expresses an opinion veiled as fact, as well as against the use of

Some students disregarded the different purposes of academic papers and encyclopedia articles. Like conventional encyclopedias, Wikipedia has three core content policies: No original research, a Neutral Point-of-View, and Verifiability. The purpose of an encyclopedia article lies in imparting knowledge, not in presenting an argument. As a result, articles should not consist of original research or personal opinion.[13] An encyclopedia is not the place to engage in debates even though it may be appropriate to describe divisions of opinion and uncertainties about the subject-matter. If opinions are relevant to a topic, their significance and origin need to be explained and documented, just like in academic writing. If views on a topic differ, these views need to be presented with impartiality. A neutral point of view means that all opinions need to be representing fairly, and as far as possible without bias.[14] As far as verifiability is concerned, Wikipedia's editorial policies insist on providing information only from reliable sources with appropriate references.[15] This means that students had to consult the traditional sources of legal scholarship, such as published textbooks and journal articles, government reports, as well as other reference works, respected sources and documents, to obtain a comprehensive and reliable picture of their assigned subject-matter.

Not all students felt equally comfortable with being required to write in a format that differed from traditional essay style assessment. One student frankly admitted after the exercise that she 'appreciated the content, but didn't like the format' of the Wikipedia assignment. Another student was more self-reflective when he described his experience in the following way:

> The wikipedia article, and particularly the review was met with a degree of cynicism at first, because it seemed very 'high school'. However once you sit down to write the article, you realise there is much more to it - it takes longer and more effort to write about a potentially complex legal topic in a way that sounds simple and is pitched at a lay audience. That was probably the most valuable learning experience.

In contrast to text-based reference works, Wikipedia as an online resource allows for multi-media context. Articles can be hyperlinked to with related articles, external links and websites as well as be illustrated with pictures, videos, diagrams and so on. Only a small number of students made use of these possibilities and provided a suitably enriched description of their topic. In some other cases, students needed to be reminded that copyright regulation also applies to the internet, including to Wikipedia.

Evaluation of the Project

The project was evaluated from the student perspective, teacher perspective, from the perspective of skills development in law students, and from the broader perspective of engaging students with a community of knowledge.

"Peacock Terms" (such as "beautiful", "well-known", "significant", and "obvious"), which contain unverifiable evaluations; see Wikipedia: Guide to Writing Better Articles, http://en.wikipedia.org/wiki/Wikipedia:Guide_to_writing_better_articles (December 4, 2009).

[13] See Wikipedia: No Original Research, http://en.wikipedia.org/wiki/Wikipedia:No_original_research (December 4, 2009).

[14] See Wikipedia: NPOV, http://en.wikipedia.org/wiki/Wikipedia:NPOV (December 4, 2009).

[15] See Wikipedia: Verifiability, http://en.wikipedia.org/wiki/Wikipedia:Verifiability (December 4, 2009).

1. Student feedback

On the whole, students reacted very favorably to this novel type of assessment. In anonymous surveys after completion of the unit, students described the assessment as 'original and thought provoking' and commended the unit for its 'diversity of assessment'. As a new form of writing, it caught students' attention. An obvious advantage of Wikipedia is that students are engaged in a 'real life exercise'. If a student sees that their writing has significance beyond the assessment, they are likely to be personally more engaged. The connection to the 'real world' is one of the hallmarks of student learning activities that have 'thick authenticity' (Schaffer, D.W. & Resnick, M., 1999).[16]

In an end-of-semester unit evaluation, the statement that the 'the assessments provided me with diverse and challenging tasks' scored a mean of 4.27 (out of 5.00) and the statement that 'the Wikipedia Assignment was an innovative and useful learning experience' scored 4.26 (out of 5.00).[17] One student made the following instructive observations after the end of the unit:

> The wikipedia article was certainly a unique assessment piece, and one particularly well suited to Comparative Law. It helped us understand snippets of an area of study with which none of us were particularly familiar.

A number of students were critical of certain aspects of the exercise. Some commented that, in light of the workload involved, the exercise should have been worth more than the allocated 20% of the final mark. Others would have preferred greater guidance on how to write for the Wikipedia assignment. These are valuable suggestions that should be borne in mind for the design of similar exercises are design for a future.

The fact that students were free to write and publish to Wikipedia, or could opt to present their Wikipedia article as a word document apparently also caused some confusion. Requiring actual publication may have been more straightforward and would also have added further authenticity to the exercise. Now that Wikipedia provides even more step-by-step information on this process there is probably little justification not to insist on actual uploading.

2. Demands on teacher

While the workload involved with the exercise was considerable, this was partly due to the fact that not only students, but also the teacher, had to become familiar with the writing process in the Wikipedia environment. Where students are likely to need much guidance, the exercise might be more suited to smaller classes. A limited group size also allows each student, or a small group of students, to work on individually assigned topics. While having a multitude of topics increases the marking load, compared to a single topic for the whole cohort, there are a number of options to facilitate the assessment process. One option is, as described above, to ask students to internally one another's review postings and take this peer feedback into account in the assessment. Another option is to nominate the article for peer review on Wikipedia.

As time progresses, and knowledge accumulates, it is bound to become more difficult for students to write new articles on areas that have not already found some treatment in

[16] Other dimensions of thick authenticity are that learning is personal, related to the discipline and aligned with assessment.

Wikipedia yet meet Wikipedia's notability standards. The collobarative and public nature of Wikipedia also stands in the way of setting substantially similar assignment in consecutive years, especially if student do indeed upload their work. This does not necessarily limit the use of Wikipedia because it may even be preferable for students who are Wikipedia beginners to critically analyze existing articles and to identify and correct any errors they find (Ayers, Matthews & Yates, 2008). In this way, Wikipedia's characteristic of not having authoritative content control can be turned into a virtue and used to educate students in critically engaging with it.

Beginners, be they students or educators, can now draw on a well-developed range of online support materials. These resources do not only cover the first steps but also contain guided tours, suggested exercises, guidelines and instructions, templates, and a help desk.[18] They include Wikiversity, an online centre which sees its mission in the creation and use of free learning materials and activities.[19]

3. Improving research and media literacy skills

The Wikipedia assignment assisted students in reflecting on the differences between various types of sources, on how they are produced and when it is appropriate to rely on one rather than another. Furthermore, the exercise helped improve students' skills to engage critically with material they found on the web and provided them with tools to evaluate the trustworthiness of information. Being involved in the production of internet content provided students with a change of perspective. Even for students who are aware that everyone can change content on Wikipedia, it is a vastly different experience to actively participate in this process and themselves create or change information available on the internet.

4. Engagement with a community of knowledge

One of the most impressive aspects of the ambitious Wikipedia project is that it aims to engage individuals from around the world to form a global community collating and disseminating knowledge. Research into Wikipedia suggests that there is often a natural progression in which Wikipedia users generally begin as seekers of information, but gradually become increasingly active with Wikipedia by proofreading, fact checking, before eventually authoring new content (Bryant, Forte & Bruckman, 2005). An assessment which requires students to become acquainted with the production of Wikipedia not only breaks down the barrier between passive and active contribution. It also gives students the opportunity to 'publish' their work on Wikipedia and contribute to the sharing of knowledge. Learning activities involving use of online collaboration enable students to feel as a part of this community of knowledge and provide 'a unique opportunity for student writers to enrich public discourse in a way that serves a real purpose and engages a real audience' (Forte & Bruckman, 2006). Contributing to this process of sharing knowledge with others improves student autonomy, their media literacy skills and encourages them to engage in a collaborative effort to create, edit and disseminate knowledge.

[17] Where 5 is "strongly agree" and 1 is "strongly disagree".

[18] See Wikipedia, School and University Projects, at http://en.wikipedia.org/wiki/Wikipedia:School_and_university_projects (December 4, 2009).

[19] See Wikiversity, http://en.wikiversity.org/wiki/Wikiversity:Main_Page (December 4, 2009).

The idea of Wikipedia as a global community of knowledge had a particular appeal in this unit. Comparative Legal Studies, by their very nature, span beyond jurisdictional borders. Comparative law not only aims to provide us with knowledge of the laws in place in foreign jurisdictions. It also enables a better understanding of domestic law by contrasting it to the solutions adopted elsewhere and providing us with potential solutions for social problems and legal issues. When lawyers from different jurisdictions come together, they may even occasionally gain insights about the universal nature of law and its development, and ultimately deepen our understanding of different societies and cultures. As one student perceptively commented:

> The Wikipedia format, being internationally 'accepted' and recognised, ties in well with the concept of studying and comparing foreign legal systems.

The aims of comparative law thus resonate particularly well with the basic premise of Wikipedia.

CONCLUSION

Wikipedia assignments are a welcome addition to the traditional writing tasks for students. Requiring students to engage actively with the content of Wikipedia, its production and limitations will enable students to become more critical and self-aware in their use of this now ubiquitous online resource. It also improves student skills in writing in a different environment and for a wider audience, while at the same time improving student engagement through a 'real world' exercise.

The skills which writing for Wikipedia requires vary, but are not in all respects different, from those practized in standard writing assignments. The most notable differences concern the purpose, format and tone. As a reference text, Wikipedia should not contain original research or debatable claims. Nonetheless, students still need to gain sound knowledge of the subject area through thorough research into the traditional scholarly materials. They will also need to continue to provide appropriate references to the sources they use and remain accountable for their quality and currency.

There are now ever more resources at hand to allow teachers unfamiliar with this assessment type to test the water and adopt it in their teaching. Students in the project discussed here responded very favorably to being set an innovative and challenging task that really engaged them.

REFERENCES

Ayers, P., Matthews, C. & Yates, B. (2008). *How Wikipedia Works* (San Francisco: No Starch Press)

Bryant, S., Forte, A. & Bruckman, A. (2005). Becoming Wikipedian: transformation of participation in a collaborative online encyclopedia, Proceedings of ACM GROUP: International Conference on Supporting Group Work, Sanibel Island, FL, 1-10.

Busemann, K. & Gscheidle, C. (2009). Web 2.0: Communitys bei jungen Nutzern beliebt, *Media Perspektiven*, 356-364.

Forte, A. & Bruckman, A. (2006). From Wikipedia to the Classroom: Exploring Online Publication and Learning, at ICLS '06: Proceedings of the 7th international conference on Learning sciences (2006), 182-188, available at http://www.cc.gatech.edu/~aforte/ForteBruckmanFromWikipedia.pdf (December 4, 2009).

Geiger, R. S. (2008). Conceptions and Misconceptions Academics Hold About Wikipedia, presentation at Wikimania conference, available at http://www.stuartgeiger.com/wordpress/conference-presentations/2008/07/18/conceptions-and-misconceptions-academics-hold-about-wikipedia/ (December 4, 2009).

Jenkins, H. (2007). What Wikipedia Can Teach Us about the New Media Literacies, Keynote Address presented at the National Media Education Conference 2007, available at http://www.henryjenkins.org/2007/06/what_wikipedia_can_teach_us_ab.html (December 4, 2009).

Schaffer, D. W. & Resnick, M. (1999). "Thick" authenticity: New media and authentic learning (1999) 10 *Journal of Interactive Learning Research.*, 195-215.

Wedemeyer, B. (2008). 'The quality of scientific articles on the English Wikipedia', presentation at Wikimania 2008 conference, available at http://www.kaltura.com/devwiki/index.php/The_quality_of_scientific_articles_on_the_English_Wikipedia (December 4, 2009).

Witzleb, N., Martiny, D., Thoelke, U. & Frericks, T. (1999). Comparative Law and the Internet (1999) 3 *Electronic Journal of Comparative Law*, available at http://law.kub.nl/ejcl/32/art32-1.html (December 4, 2009).

In: Computer-Assisted Teaching: New Developments ISBN: 978-1-60876-855-4
Editors: B. A. Morris, G. M. Ferguson, pp. 201-211

Chapter 11

AFFORDANCES OF "MY LEARNING SPACE" FOR CONSTRUCTIVIST LEARNING

***Nadarajah Loganathan*[*] *and Qiyun Wang*[**]**
Learning Sciences and Technologies Academic Group, National Institute of Education, Nanyang Technological University, Singapore.

ABSTRACT

Information and communication technology (ICT) has been popularly used not only in schools but also in various organizations like training centers. They do not just simply use ICT tools to present content or let students or trainees do drill-and-practice. They are now more using an integrated learning management system to conduct and facilitate the whole teaching or training process. In this chapter, we present the basic features and pedagogical, social, and technological affordances of a learning management system (called my learning space) which is used by the Signal Institute of the Singapore Armed Forces (SAF). The perceptions of trainees on the use of the learning space are also briefly descried.

INTRODUCTION

In the transformation of the 3rd Generation Army, the SAF needs to produce knowledge-enabled soldiers and leaders. The learning environment used to train soldiers needs to be more learner-centric and focus on cognitive, affective and psychomotor skills which are aligned to the vision of "Learning Army, Thinking Soldiers". Moreover, as the global environment requires soldiers to master more competencies within a limited amount of time, it is imperative that the learning environment be transformed to adapt to the current profile of people who are more ICT savvy. In the pursuit of the "Learning Army, Thinking Soldiers" vision, the SAF faces the need to update existing training methodology and fuses it with the

---

[*] Corresponding authors: Email: logan_mba@yahoo.com.sg
[**] Email: Qiyun.wang@nie.edu.sg

rapid technological advances in the ICT arena. This will eventually allow soldiers to learn how to construct knowledge on their own, and acquire vast relevant competencies in the shortest time.

For this purpose, the Signal Institute spearheads the promulgation of ICT for the SAF, developed the "My Learning Space (MLS)" environment. Its intended audience would be the new generation of soldiers who have already been exposed to new learning technologies in the current information technology landscape. The focus of the environment is to maximize self directed learning through the development of a customized learning environment, and to transform the traditional "instructor-centric" teaching into more learner-centered.

This chapter provides an overview of the MLS as a tool that has been used to support e-learning. The overall aim of the MLS is to enhance the ways in which ICT can be used in Military training. The primary idea of this tool is to give soldiers the flexibility to learn at their own pace. This paper presents the instructional ingredients of the system from pedagogical, social and technological perspectives.

The MLS is a web-based system, developed in-house using PHP, Flash and SQL, so as to facilitate deployment as a standalone mode as well as on the Intranet and Internet. Figure 1 is a screen capture of the MLS. This system has six integrated modules:

a. *e-Training Program Module*. This module is designed to allow trainees to access the training program, which is updated by the Training Resource Section in Signal Institute. Related links will point to Lectora courseware, "How-to" videos in the ShowMe Module and lecture videos in the Digiteach Module.
b. *AskSME Module*. This module is to allow trainees to familiarize themselves with the various subject matter experts (SMEs) and also to interact with the SMEs via instant messaging. This is meant to augment the traditional approach of meeting people face-to-face. Information displayed in this module includes the appointment history, areas of expertise (eg. High Frequency Radios), areas of experience (eg. Involvement in overseas missions or major overseas exercises) and courses attended (eg. Certified CISCO Network Academy 1).
c. *ShowMe Module*. This module is designed to display videos in practical lessons. This mode of learning allows for repeated viewing of the same video clip.
d. *Digiteach Module*. This module captures all lectures on videos and integrates with the presentation slides. This module allows trainees to view with synchronised slides.
e. *Action Learning Process Module*. This module is intended to allow trainees to learn from the before, during and post action reviews and also to input new data. Three modes to display the information from the reviews are available, which are the "Forum" mode, "Mindmap" mode, and "video" mode.
f. *Online Assessment Module*. Trainees are able to take tests online to assess their understanding of various topics and also track the progress. Instructors are also able to view the progress of the trainees and pay closer attention to the trainees.

Pedagogical, Social and Technological Perspectives

Numerous instructional models can be adopted to integrate ICT into a curriculum. Each of these models has different considerations or learning orientations. Nonetheless, one of the most basic models was developed by Wang (2008). This model comprises three basic components: pedagogy, social interaction and technology as shown in Figure 2.

The term of pedagogy, in this model, refers to teaching strategies, techniques or approaches that teachers adopt to deliver instruction or facilitate learning. This chapter shows how the digital technology can be harnessed in transforming the way we impart military knowledge. A myriad of teaching strategies encompassing videos, Power Point slides, forums have been incorporated into the MLS. This space has adopted a constructive approach for the 3rd Generation transformative learning. The MLS ensures that it deals with individual differences and incorporates practices to optimize learning.

The term social interaction refers to two-way communication. It can be student-teacher or student-student interaction. The MLS is developed to promote collaborative learning. The idea of collaboration in the army is to foster creativity and self-fulfillment, and to create new values for soldiers by encouraging them to drive their own conclusions collectively. Technology is defined as a supporting tool. It is therefore inevitable for Signal Institute to harness the appropriate emerging and current technologies into the MLS.

We now see the affordances of the pedagogical, social and technological components in the MLS. If we take the analogy of a road system, then a good network of a road would form the most basic condition for good transport services. Next, we need reliable and roadworthy cars and buses to ply the road to ensure it does not get congested by broken down vehicles, and lastly, we also need well trained drivers to keep the road safe. In the same context, effective integration of ICT would be the most fundamental and critical component in a technology-enhanced learning environment. This should be augmented by sound design of pedagogy and reinforced with an environment that promotes social interaction. Without sufficient support of technology, undoubtedly many pedagogical and social design activities, such as synchronous or asynchronous online discussions, would be hard to implement. On the other hand, the primary factor that influences the effectiveness of learning is not the availability of technology, but the pedagogical design and social design (Mandell, Sorge, & Russell, 2002).

Figure 1. A screen capture of the My Learning Space.

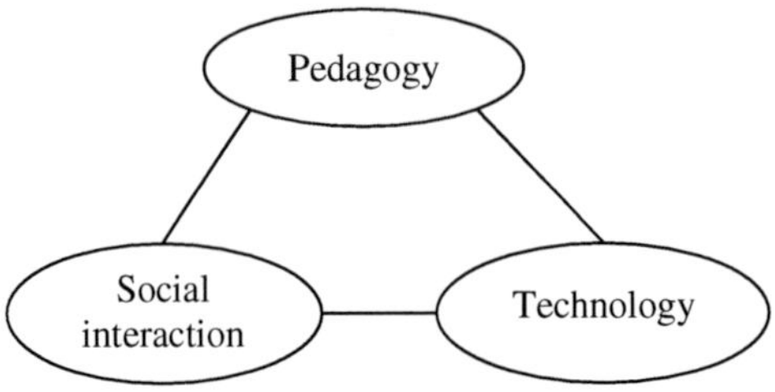

Figure 2. Key components of the model.

AFFORDANCES

Affordances are about effectivity. The word "affordance" was originally invented by the psychologist Gibson (1977, 1979) to refer to the actionable properties between the world and an actor (a person or animal). To Gibson, affordances are relationships. They are a part of nature: they do not have to be visible, known, or desirable. Affordances, according to Gibson, should not be further viewed as dependant on culture, prior knowledge or expectations of the individual. By disregarding knowledge and expectations of the actor, Gibson focuses only on the action capabilities of the actor and not on the perceptual and mental capabilities. However, the information that specifies the affordance is indeed dependant on the actor's experience and culture.

Here, we define affordance as what an object suggests to us. It is a combination of perception and cognitive constructs. For example, if you see a bench, you might think to sit down on it, or to lie down on it. Some doors have a panel on one side, and a handle on the other. If you see the panel, you would think to push it. If you see the handle, you would think to pull it. We would use the term affordance to talk about how the features in the MLS allow trainees to perceive and develop their mental construct from the contents provided to enhance their learning. The following presents the affordances offered by the three components; pedagogy, social interaction and technology.

Pedagogical Affordances

In the constructivist learning theory, the emphasis is placed on the learner or the student rather than the teacher or the instructor. It is the learner who interacts with contents and events and thereby gains an understanding of the features held by such contents or events. The learner, therefore, constructs his/her own conceptualizations and solutions to problems. Students learn best when they actively construct their own understanding.

The 3rd Generation Army intends to develop competent, confident and committed soldiers. This expectation has fundamentally caused a behavioral change from both the trainees and trainers. The 3rd Generation soldiers though come from a diverse background, certainly are very ICT savvy. Besides, they are the post 80s who are independent and self motivated. They need little supervision. Having the autonomy to decide on their learning modes and leveraging on their self motivation, a majority of our trainees not only learn the "whats" and "hows" of the contents, but also want to know the "whys". In terms of pedagogical design, the MLS environment supports and satisfies the needs and learning intentions of trainees in the following areas:

a. *Abundance of Resources.* The MLS has several resources for trainees to acquire knowledge. In the past we used to hand over précis to trainees on the very first day they report to the institute. Today, the instructor can upload the soft copy of these précis into a server for trainees to access even when they are in the comfort of their homes. The trainees are given access to both the internet and intranet to acquire knowledge beyond textbooks / précis. Both the intranet and internet provide trainees access to different information resources that help them understand the content better.
b. *Knowledge Management.* A networked writing program provides a unique platform for collaborative writing. When trainees come together to discuss their experiences, the program allows them to share what they have learnt and also to facilitate group learning. This knowledge is collectively documented in text, to allow the following batch of trainees to learn from this present group's experiences thus making their learning curve shorter and steeper.
c. *Access to SMEs.* Telecommunication tools such as e-mail and the Internet provides the means for dialogue, discussion, and debate. These promote interactivity and social knowledge construction. Using the "AskSME" Module allows trainees to familiarize themselves with the various subject matter expert and also to interact with them. Figure 3 is a screen capture of the AskSME module.
d. *Meeting different learning styles.* Different trainees have different learning styles. Based on the book 'How to Maximize Your Child's Learning Ability' by Dr. Lauren Bradway and Barbara Albers Hill, the authors come to the conclusion that children have distinctive learning styles. They believe that a child's learning style in his infancy influences his ability to socialize, and to perform athletically and academically later in life. Every child has a specific way of absorbing information. Some learn best through visual stimulations; others are more responsive to sound and language; while some learn more quickly through touch and motion. Therefore, they broadly classified children into three categories based on their preferred learning styles: lookers, listeners, and movers. In the MLS, apart from looking at the 'Digiteach' module which captures all lectures on videos and integrate with the presentation slides, the 'ShowMe' module is designed to display videos on practical lessons. Figure 4 is a screen capture of the ShowMe module. In the near future, the Institute plans to upload movies like "Battle of Midway" to illustrate the importance of radio discipline in wars. There is a big scope for meaningful learning using this module as the content becomes richer. These initiatives are undertaken with the intent of meeting the needs of the learners according to their learning styles.
e. *Developing Critical Thinking Skills.* In the MLS environment, the instructors are facilitators or guides. These instructors provide scaffoldings to trainees. Trainees come out with their own reflections. These reflections help them to develop their analytical and critical thinking skills as the MLS is designed not to teach the trainees but to elicit the information from the trainees themselves.

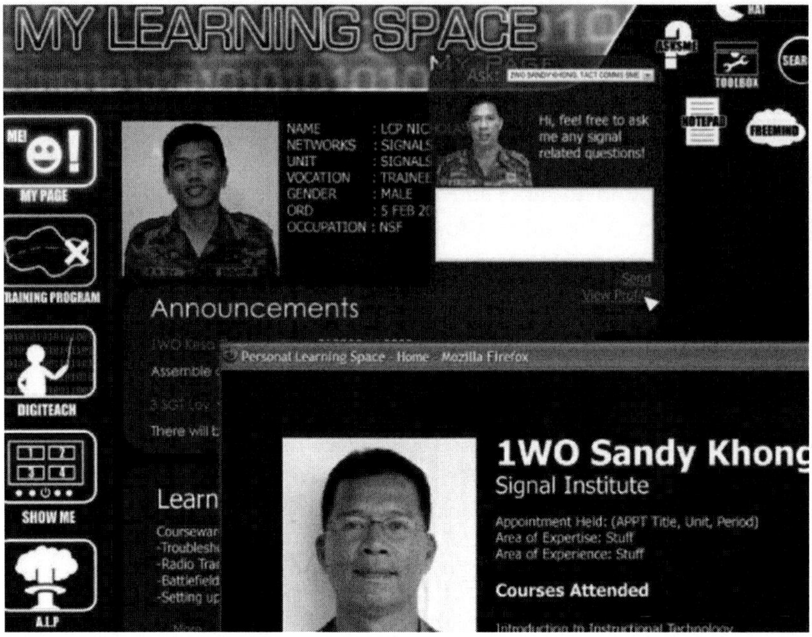

Figure 3. A screen capture of the AskSME module.

Figure 4. A screen capture of the ShowMe module.

The pedagogical affordances of the MLS allow students to learn by fitting new information together with what they already know, and hence support the constructivist theory. An additional fact about the MLS is that the system also supports traditional teaching in that the content design of each lesson adheres very closely to the nine events of Robert Gagne's theory of instructions. For instance, the intent of allowing trainees to access the training program is to gain the attention of the trainees. Each courseware / video highlights the "Objectives" of the courseware or the lesson clearly. If the courseware / video is linked to the previous lesson, a quick revision is provided not only to refresh trainees' knowledge, but also more importantly to ensure that the trainees have the necessary knowledge to progress.

The courseware / videos are designed such that some startling fact, a short story, a rhetorical question or a historical illustration is given to both gain the attention and also elicit performance. At the end of the lesson, all trainees would have to complete an online assessment. Instructors are able to view the progress of the trainees and provide coaching to weaker trainees. All trainees are also expected to give their feedback on the module after each lesson, so that the designers of the module can improve on the software / video. All these instructional events are adhered to ensure that the trainees are able to achieve the predetermined learning outcomes.

Social Affordances

The MLS also has certain social affordances. The "AskSME" module allows for users to search the database on topics or instructors. This is meant to facilitate social networking by incorporating data on subject matter experts and their profiles. This will enable faster peers learning by drawing on their experiences from the wider pool of experts. Secondly, it enhances collaborative learning. When trainees come together to discuss their experiences, it allows them to share what they have learnt and also to facilitate group learning. This knowledge can be captured either in text or mind maps videos. Trainees can then view the documented knowledge in order to better prepare themselves for future exercises. This collaborative learning process has shown positive results as trainees are not only quick to apply what they had learnt during the field environment, but also able to identify potential pitfalls even before they are put into the field environment.

The availability of both synchronous and asynchronous modes in the system is another support of social affordance. Trainees have the flexibility to log on at any time, any place to post questions or assess the sources available in the MLS. Likewise, the instructors have the latitude of time to address queries posted by trainees. Unlike, in the past, if the trainees had any questions for the trainer, they would have to wait to catch the instructor along the corridor or till the instructor comes for the next lesson to have their queries answered. What mattered most to the trainees is the comfort they feel when they post queries on the forum as they are very certain that someone will read their postings and that they will be replied soon.

Technological Affordances

Learning is always distributed in some forms among the technology, the learner and the context. The MLS is a web-based tool using PHP, Flash and SQL. It incorporates some software tools like Skype, windows media player and Youtube. By incorporating these software tools, the MLS provides an avenue for "Learning on Demand" besides offering a variety of teaching methods. Though some of these features have been highlighted earlier, the crust of the matter is that these features are only possible when technology has advanced to meet the aspirations of instructional designers. Technology today affords us the following features in MLS;

a. *Learning through videos.* There is great potential for practical lessons as videos will help to reinforce the learning process and facilitate revision of past lessons. The customised module allows for users to skip to specific sections of the video to focus on the frames of interest. Instructors will be empowered with the capacity of adding values to their lessons. Figure 5 shows a screen capture of a lecture in a video.
b. *Learning on Demand.* The Digiteach Module is another mode for the Signal vocationist to update themselves as and when required in the process of continuous learning. Regular Signal vocationists can view the lecture videos to familiarize themselves with the new subjects.
c. *Access Control.* The Signal Institute conducts an array of courses within the institute. The students participating in these courses are from very different background and also have varying demands on the access of information. The students from the basic courses are generally not required to have any access to 'Classified' information, whereas students attending the advance courses require access to information from the intranet. It would also be helpful for them to be able to access the internet for their projects. For this purpose, the MLS is designed to meet these requirements.
d. *Facilitate conduct of E-exams.* The MLS is also designed to be standalone. This feature allows our students to use the laptops for the examination. A day prior to the exams, students would surrender their laptops for the trainers to upload all the necessary resources and questions into the laptops. After the exams, all the students' submission is saved into the tutor's computer for the convenience of the markers.

An online learning environment does not preclude that it is an e-learning environment or an excellent e-learning tool. It is crucial for a good e-learning tool to have a good interface design as it determines the usability of the technology-based learning environment. The interface design of a computer program ought to focus on ease of learning, ease of use and aesthetics (Wang & Cheung, 2003). Ease of learning is critical for beginners while ease of use become more important while users gain experience over time. Certainly, the interface must be more attractive so that it can motivate and engage learners.

Figure 5. A screen capture of a 'Lecture' video.

Table 1. My Learning Space Experience.

	Positive	Negative	Neutral
Perception of MLS a good tool for effective learning (P)	148	8	20
Prior knowledge on e-learning tools (P)	168	8	-
Sufficient time provided for research (P)	166	-	10
Active participation by all trainees in all group and individual projects (S)	170	1	5
Information relevant to the course was more from internet/ intranet source (P)	151	6	19
Instructor was always supervising us (P)	20	151	5
Knowledge resources are abundant (P)	176	-	-
The interface design for MLS is user friendly (T)	160	8	8
Relevance of content and applicability of modules (P)	172	-	4

Enhancing "My Learning Space"

A survey was conducted at the end of the training with the trainees who were in the pilot program. The primary objective of the survey was to determine if the trainees found the MLS as an effective tool for learning. In the process, we also intended to know how they felt about the affordances in the pedagogical, social interaction and technological aspect of the MLS.

Of the 176 students surveyed in Dec 2008, 148 students reflected that the MLS was perceived as a good learning tool as it gave them the flexibility to learn at their own pace and place. 20 students were neutral while 8 felt they preferred lectures and proper lessons as they felt that their 'free time' was taken away by the project work. Some of the key results of the other questionnaires in the surveys are listed in Table 1.

Although the survey showed very good ratings for all the three affordances, there were additional feedbacks from the students to further enhance the social affordance in the MLS. For instance, the current system allows only instructors to upload photographs onto the system for trainees to relate the name to the face. Most trainees felt that the environment could be made more 'friendly' if trainees could post their own photographs onto the MLS, so that instructors, who now see much less of them, can relate the name to the face when doing discussions on cyberspace. For that particular cohort, they circumvented the problem by inviting their friends and instructors to Facebook. On the other hand, the instructors themselves did not feel that it was a major issue as all instructors had a 'pen-picture' of the trainees with them. Moreover, the instructors claimed that the MLS had reduced only forty percent of their face to face instructional time with the trainees. The instructors were certain that they would be familiar with their trainees' faces within the first week of the course.

Another feedback from the trainees was for the trainer not to forsake the précis and handouts issued to them. This cohort of trainees was not issued with any hard copy précis as opposed to other cohorts. Trainees felt that though they had the autonomy to decide on the management of the laptops, they seldom wanted to bring it home. This was primarily because they found it cumbersome to carry an additional laptop home. However, they reflected that having hard copy of the précis gave them the flexibility to read them on the journey either to or from camp.

CONCLUSION

The MLS has the key features to meet the learning needs of the next generation soldiers. The functions available in the MLS allow soldiers to operate on a self directed learning mode in a customized learning environment. While the utilities meet the learners' needs, the technological affordances offered in the MLS are user friendly. In the current state, based on the students' feedback, trainers need to learn how to become more efficient facilitators to make online discussions more effective. While the MLS has certain affordances, we would continue to explore how to further improve the system from the pedagogical, social and technological aspects.

REFERENCES

Gibson, J. J. (1977). The theory of affordances. In R. Shaw & J. Bransford (eds.), *Perceiving, acting and knowing*. Hillsdale, NJ: Erlbaum.

Gibson, J. J. (1979). *The ecological approach to visual perception*. Boston: Houghton Mifflin.

Mandell, S., Sorge, D. H. & Rusell, J. D. (2002). Tips for technology integration. *Tech Trends, 46*(5), 39-43.

Wang, Q. Y. (2008). A generic model for guiding the integration of ICT into teaching and learning. *Innovations in Education and Teaching International*, *45*(4), 411-419.

Wang, Q. Y. & Cheung, W. S. (2003). Designing hypermedia learning environments. In S. C. Tan, & F. L. Wong (Eds), *Teaching and learning with technology: An Asia-pacific perspective* (pp. 216-231). Singapore Prentice Hall.

INDEX

A

academic performance, 65, 66
academics, 191, 192, 200
acceleration, 47, 53
accessibility, 171
accounting, 82
accreditation, 176
accuracy, 22, 80, 82, 128, 174, 192
acetate, 40
achievement, 63, 66, 110, 112, 115, 137, 144, 146, 147, 193
acid, 11, 12, 19
ACM, 165, 188, 199
action research, 97, 98, 111
activation, 17
acute, 173
adaptability, 130
adaptation, 9, 59, 127
adaptive control, 132
adaptive multimedia, 117, 128, 130
administration, 95, 106, 144
administrators, 108, 109, 179
adolescents, 89
adult, 46, 101, 103, 113
adulthood, 113
adults, 46, 96, 188
aerobics, 135
aesthetics, 208
Africa, 97, 111, 171
age, 42, 45, 49, 55, 58, 111, 136, 138, 145, 179, 192
agents, 65
agriculture, 180
aid, 53, 121, 161, 176
air, 52, 127, 129, 131
alchemy, 33
alpha, 47, 56, 98, 104, 140, 141
alternative, 11, 37, 43, 44, 45, 59, 60, 107, 120, 121, 126, 127, 186, 193
alternatives, 103, 115, 127
alters, 51
amendments, 194
American Educational Research Association, 90, 91, 111, 114, 131
ammonia, 107
amplitude, 51
Amsterdam, 35, 130, 132, 162
analysis of variance, 133, 140
anastomosis, 175
anatomy, 167, 173, 175
animal models, 175, 176
animals, 174, 177
animations, 4, 5, 7, 9, 14, 18, 19, 20, 21, 29, 122, 123, 129, 131
annotation, 160
ANOVA, 106, 133, 140, 142
antecedents, 112
anthropometry, 113
Archimedes, 77, 83
Aristotle, 114
Armed Forces, 201
Army, 201, 204
artificial intelligence, 128, 152
Asia, 171, 211
assessment, 1, 4, 10, 18, 19, 20, 21, 29, 33, 69, 71, 78, 100, 101, 128, 129, 130, 132, 172, 191, 192, 193, 194, 196, 197, 198, 199, 207
assignment, 194, 196, 197, 198

assumptions, 9
asynchronous, 4, 10, 25, 27, 29, 30, 153, 155, 163, 186, 203, 207
Athens, 147
atmosphere, 30, 56
attacks, 182
attitudes, 31, 97, 110, 113, 114, 115, 127, 133, 135, 136, 140, 161, 188
Australia, 93, 97, 99, 111, 112, 113, 114, 117, 191
Austria, 151
authenticity, 33, 41, 70, 197, 200
automation, 7, 168
autonomy, 56, 198, 204, 210
availability, 109, 110, 156, 175, 193, 203, 207

B

back, 39, 42, 51, 52, 100, 108, 156, 180, 182, 187
barriers, 39, 57, 60, 198
basketball, 135, 143, 144, 146
behavior, 102, 111, 117, 118, 128, 161
behavioral change, 204
beliefs, 98, 111, 113
benefits, 7, 30, 95, 108, 125, 160, 179, 182, 188, 191
bias, 196
binding, 195
biosensors, 180
blocks, 101
blogs, 156, 181, 188
Bluetooth, 175
Boston, 32, 35, 113, 165, 210
bowel, 175
boys, 136
brainstorming, 75, 157
broad spectrum, 167
broadband, 134, 175, 180
browsing, 91, 129, 182
bubble, 79
buffer, 19
buses, 203
buttons, 53, 137

C

calibration, 19, 22, 24
Canada, 92, 113, 114, 163, 171, 179, 181
candidates, 110
case study, 33, 57, 114
CBS, 98
cell phones, 188
certification, 171, 174, 177
channels, 7, 8, 118, 154, 155, 157
cheating, 188
childhood, 114, 185
children, 39, 43, 44, 57, 60, 61, 97, 134, 135, 205
cholecystectomies, 176
cholecystectomy, 178
chunking, 123
class period, 138
class teaching, 42
classes, 10, 18, 39, 44, 50, 55, 96, 97, 98, 100, 101, 103, 105, 106, 108, 111, 114, 135, 145, 163, 172, 179, 180, 182, 183, 185, 197
classroom environment, 94, 95, 96, 97, 106, 110, 111, 112, 113, 114, 115
classroom practice, 92, 93, 98, 192
classroom settings, 186
classroom teacher, 37, 38, 43, 95
classroom teachers, 38, 43, 95
classrooms, 38, 39, 57, 63, 64, 65, 93, 94, 95, 96, 97, 98, 99, 100, 101, 106, 107, 108, 109, 110, 112, 114, 115, 158, 160, 179, 180, 181, 182, 187, 188
clinics, 177
CME, 171, 175
codes, 193
coding, 130, 131
coffee, 180, 188
cognition, 59, 60, 111
cognitive capacity, 8, 119
cognitive development, 61, 68
cognitive flexibility, 132
cognitive function, 118
cognitive level, 68
cognitive load, 2, 6, 7, 8, 18, 21, 25, 28, 32, 33, 34, 35, 49, 80, 117, 118, 119, 120, 121, 122, 123, 124, 125, 126, 127, 129, 130, 131
cognitive process, 7, 31, 67, 68, 70, 78, 80, 87, 91, 118, 124
cognitive processing, 7, 124
cognitive science, 45, 60, 131
cognitive system, 7, 118
cognitive tool, 60, 66
cohort, 110, 197, 210
collaboration, 6, 9, 10, 21, 24, 25, 29, 30, 31, 68, 93, 106, 109, 152, 154, 157, 158, 161, 162, 164, 184, 186, 198, 203
Collaborative learning, 29, 71
collateral damage, 173
college students, 66

colleges, 38, 46, 47, 49, 182, 183
collisions, 33
communication, 4, 18, 19, 21, 37, 38, 60, 63, 64, 66, 68, 152, 153, 154, 155, 156, 157, 160, 161, 172, 174, 180, 183, 184, 201, 203
communities, 92, 108, 154, 155, 164, 172, 184
community, 63, 66, 102, 108, 154, 155, 156, 160, 162, 167, 172, 176, 185, 186, 196, 198, 199
compatibility, 194
competence, 88, 152
competition, 108, 149, 184
complex systems, 89, 122
complexity, 6, 28, 44, 91, 119
complications, 139
components, 53, 88, 98, 119, 120, 121, 126, 203, 204
composition, 21
comprehension, 8, 91, 125, 129, 131
compulsory education, 46
computational modeling, 128
computer simulations, 5, 33
computer skills, 145
computer software, 42
computer technology, 134, 167, 173, 176
computing, 112, 163, 180, 182
conceptualizations, 204
conductivity, 56
confidence, 20, 38
configuration, 20
confusion, 197
congruence, 96, 122
connectivity, 172
consensus, 9
consent, 136
constraints, 130, 168, 172
construction, 7, 22, 45, 64, 66, 67, 71, 79, 80, 83, 86, 87, 124, 125, 138, 159, 174, 205
constructivist, 59, 68, 71, 90, 93, 94, 95, 96, 97, 98, 99, 100, 101, 103, 106, 107, 109, 110, 111, 113, 114, 204, 207
constructivist learning, 68, 97, 100, 101, 109, 111, 114, 204
consultants, 138
consumers, 155, 192
contamination, 181
continuity, 94, 177
control, 5, 8, 21, 64, 78, 81, 84, 85, 86, 96, 101, 106, 111, 128, 129, 131, 132, 143, 145, 157, 175, 198
control condition, 120, 127
control group, 64, 81, 84, 85, 96, 143
conversion, 30
cooperative learning, 68, 151, 152, 153, 154, 155, 156, 157, 160, 161, 162
correlation, 1, 16, 17, 126, 141
correlation coefficient, 141
correlations, 16, 17, 126
cost-effective, 171
costs, 5, 171
Coulomb, 47
course content, 137
covering, 42, 133, 145
creativity, 68, 203
credit, 171, 176
critical points, 45
critical thinking, 2, 21, 68, 154, 205
criticism, 3, 192, 195
cues, 137, 184
culture, 113, 182, 187, 204
currency, 199
curriculum, 2, 17, 20, 25, 33, 39, 43, 44, 57, 59, 68, 80, 83, 88, 89, 94, 95, 97, 98, 107, 108, 109, 111, 135, 168, 202
curve-fitting, 23
cybercrime, 181
cyberspace, 64, 210
cyclotron, 47
cynicism, 196
Cyprus, 91

D

Dallas, 93, 110
danger, 38, 58
data analysis, 19, 113
data collection, 3, 5, 37, 41, 106
data set, 183
database, 96, 160, 186, 207
debates, 170, 179, 196
decision making, 48
decisions, 12, 17, 42, 48, 57, 72, 78, 127
delivery, 10, 11, 30, 93, 107, 110, 134
democracy, 9
demographics, 108, 110
Department of Education, 92
dependent variable, 94, 96, 140
designers, 53, 57, 117, 207
developmental psychology, 152
deviation, 103, 106
Different Instruction, 147
diffraction, 56

diffusion, 171
digital cameras, 41
directives, 107
discipline, 176, 195, 197, 205
discourse, 198
discrimination, 140, 141
displacement, 51
dissatisfaction, 29, 30
disseminate, 198
distance education, 155
distraction, 182
diversity, 9, 186, 187, 197
doctors, 44
durability, 87
duration, 98, 109, 118
duties, 106
dynamic systems, 5

E

earth, 97
ecological, 210
ecology, 106
educational aims, 37, 38, 57, 191
educational assessment, 91
educational institutions, 1
educational practices, 161
educational process, 177
educational programs, 97
educational software, 43, 45, 55, 57, 58
educational system, 94
educators, 63, 64, 65, 66, 108, 134, 192, 198
elaboration, 68, 75, 76
e-learning, 9, 11, 12, 14, 15, 16, 30, 33, 130, 151, 156, 164, 170, 202, 208, 209
election, 73
electricity, 46, 56
electrochemistry, 19
electromagnetic, 43
electron diffraction, 47
electronic communications, 184
electronic materials, 1
elementary school, 108
elementary teachers, 98, 113
email, 56, 183, 184
e-mail, 4, 19, 27, 28, 205
emotion, 188
emotional, 95, 102
empowered, 208
encoding, 124
enculturation, 110
endocrine, 171
endoscopy, 176
energy, 51, 52, 53, 54
engagement, 9, 52, 58, 65, 109, 145, 162, 199
England, 38, 58, 59
enthusiasm, 17, 57, 109, 195
environmental contamination, 181
epistemology, 59, 98
equity, 110
ERIC, 92, 131
Europe, 171
European Commission, 153
evolution, 89, 151, 168, 169, 177
examinations, 46, 49
execution, 78, 137
exercise, 26, 100, 137, 174, 191, 192, 193, 194, 196, 197, 198, 199
experimental condition, 12, 120
experimental design, 81, 139
expertise, 92, 94, 95, 110, 117, 118, 119, 120, 121, 122, 123, 125, 126, 127, 128, 130, 168, 172, 175, 176, 202

F

Facebook, 153, 155, 182, 185, 210
face-to-face interaction, 1, 30, 154
facilitators, 205, 210
factor analysis, 101
factorial, 97, 98, 104, 140
failure, 65, 87, 102
faith, 179
family, 186
fear, 156, 192
fee, 180
feedback, 1, 4, 10, 18, 19, 20, 21, 28, 29, 30, 42, 44, 45, 52, 70, 80, 97, 98, 137, 138, 153, 156, 157, 183, 184, 185, 187, 194, 195, 197, 207, 210
feelings, 155
females, 98
fiber optics, 172
fidelity, 122, 171
field-tested, 110
first-time, 145
fitness, 144
flexibility, 4, 21, 28, 29, 37, 186, 202, 207, 209, 210
flotation, 77
flow, 157
focus groups, 46

focusing, 38, 82, 95, 182, 187
Ford, 120, 132
Fort Worth, 111, 112, 114
Foucault, 50
France, 167, 170, 176
free education, 46
freedom, 14, 15
friction, 52
frustration, 3, 17
fulfillment, 203
funding, 55
funds, 2

G

games, 133, 149, 182
gamma radiation, 47, 56
garbage, 186
gastroenterologist, 176
gastrointestinal, 171, 176
generalization, 100, 110
generation, 134, 151, 182, 187, 191, 202
Germany, 172, 175
gifted, 42, 55, 56, 60
girls, 135, 136
global education, 175
goals, 30, 80, 102, 106, 119, 124, 156
google, 158
government, 38, 45, 46, 55, 196
grades, 46, 114, 135
graduate students, 25
grain, 60, 99
graph, 26, 27, 41
gravitational field, 47, 52, 56
gravitational potential energy, 52, 53
gravity, 82
Greece, 136
group activities, 9, 109
group size, 197
group work, 152, 153, 155, 156, 184, 185
grouping, 108
groups, 11, 12, 14, 15, 16, 18, 25, 27, 79, 84, 85, 86, 98, 126, 133, 135, 136, 139, 140, 141, 142, 143, 144, 145, 153, 154, 157, 161, 176, 183
growth, 95, 161
guardian, 136
guessing, 50, 138
guidance, 46, 63, 66, 72, 87, 118, 119, 121, 125, 126, 127, 128, 129, 152, 175, 191, 192, 194, 197
guidelines, 44, 70, 72, 88, 171, 183, 194, 195, 198

H

half-life, 47
handheld devices, 157
handling, 2, 174
hands, 40, 49, 94, 106, 108, 109, 148, 167, 168, 174, 175, 176
Harvard, 165
health, 144, 171
healthcare, 171, 180
hearing, 8
heat, 22
height, 53
heterogeneous, 42
high school, 96, 97, 111, 113, 135, 196
higher education, 32, 46, 151, 153, 154, 155, 156, 158, 159, 160, 161, 163
high-frequency, 184
highlighted words, 4
high-speed, 172
high-tech, 40, 109
homogenous, 108
horizon, 39, 167
hospitals, 176
household, 109
human, 5, 6, 7, 21, 41, 42, 43, 58, 65, 89, 92, 130, 132, 174, 176, 178, 181
human information processing, 132
Human Kinetics, 146, 147
hypermedia, 63, 65, 66, 67, 89, 117, 118, 124, 125, 127, 128, 129, 131, 132, 146, 211
hypertext, 67, 80, 87, 89, 90, 124, 125, 129, 131, 132, 139
hypothesis, 3, 17, 30, 65, 120, 130

I

ICT, 4, 28, 32, 33, 37, 38, 39, 42, 45, 57, 58, 60, 91, 93, 96, 106, 110, 180, 185, 186, 201, 202, 203, 204, 210
identity, 107, 189
idiosyncratic, 44
images, 4, 5, 7, 8, 9, 19, 30, 35, 41, 49, 122, 137, 167, 172, 173, 174
imagination, 31
imaging, 173
immersion, 173

implementation, 9, 31, 59, 65, 68, 93, 94, 95, 98, 100, 101, 103, 106, 107, 109, 110, 111, 128, 152, 161
in situ, 132
income, 107
increased access, 184
independent variable, 84, 140
Indiana, 90
indicators, 106, 126, 138
individual differences, 95, 121, 203
individual students, 57, 99
individuality, 156
individualization, 127
induction, 58
ineffectiveness, 122
infancy, 205
Information Age, 59, 189
information and communication technologies, 37, 38, 63, 64, 180
information and communication technology, 63, 66
information processing, 6
information systems, 32, 112, 163
information technology, 1, 94, 95, 113, 163, 202
Information Technology, 147
informed consent, 136
infrared, 25
innovation, 38, 40, 96, 163, 180, 181
insight, 63, 66, 93, 99, 179, 181
institutionalization, 180
institutions, 1, 81, 180, 185
instructional design, 7, 35, 45, 64, 86, 91, 106, 118, 128, 130, 152, 208
instructional materials, 64, 66, 117
instructional methods, 118, 134, 136, 144, 145
instructional procedures, 128
instructional techniques, 117, 125
instructional time, 133, 210
instructors, 4, 31, 99, 100, 103, 106, 107, 109, 179, 180, 181, 182, 184, 185, 187, 188, 205, 207, 210
instruments, 4, 5, 22, 24, 32, 112, 128, 141, 177, 178
insurance, 46
integration, 1, 6, 9, 12, 15, 28, 31, 33, 34, 59, 65, 90, 91, 94, 95, 109, 112, 119, 120, 125, 180, 186, 187, 203, 210
Intel, 137
intelligence, 123
intentions, 111, 182, 205
interaction, 9, 29, 30, 33, 71, 123, 142, 152, 153, 157, 173, 183, 184, 203, 204, 209
interactions, 1, 9, 10, 24, 29, 30, 98, 102, 117, 119, 123, 128, 184
interactivity, 5, 17, 28, 30, 38, 40, 42, 43, 52, 55, 57, 58, 119, 134, 156, 171, 172, 205
interdependence, 9
interdisciplinary, 107, 110
interface, 4, 11, 12, 14, 16, 17, 24, 28, 42, 173, 208, 209
internal consistency, 97, 99, 101, 104, 140
internet, 40, 67, 68, 69, 70, 92, 96, 191, 192, 193, 196, 198, 205, 208, 209
Internet, 32, 35, 39, 63, 64, 65, 66, 68, 80, 87, 88, 89, 90, 91, 92, 134, 151, 152, 167, 168, 170, 171, 176, 177, 180, 186, 191, 192, 193, 200, 202, 205
interrelationships, 124
intervention, 71, 81, 84, 85, 86, 89, 108, 110, 135, 146
intervention strategies, 108
interviews, 49, 98, 99, 170
intrinsic, 6, 7, 119, 120
invasive, 167, 170, 177, 178
investment, 38
IR spectra, 25, 26, 27
IR spectroscopy, 25
isolation, 120
iteration, 193

J

Japanese, 171
Java, 156, 164
Jerome Bruner, 45
judge, 49, 138
junior high school, 97, 135, 147
jurisdictions, 199
justification, 197

K

Kenya, 146
kinematics, 32, 126
kinetic energy, 51, 52, 53
knowledge acquisition, 90, 133, 143, 145
knowledge construction, 64, 66, 67, 71, 79, 80, 205
Korea, 113
Korean, 97

L

lambda, 56, 106
landscapes, 115, 176
language, 60, 73, 97, 112, 119, 154, 158, 193, 205
laparoscopic, 168, 176, 178
laparoscopic surgery, 168
laptop, 109, 162, 173, 180, 181, 189, 210
laser, 56
law, 3, 47, 122, 192, 193, 196, 199, 200
laws, 43, 55, 77, 199
lawyers, 199
learner progress, 128
learning activity, 77, 78, 162
learning culture, 161
learning efficiency, 187
learning outcomes, 18, 33, 121, 124, 127, 140, 207
learning process, 2, 3, 5, 7, 11, 12, 17, 21, 31, 32, 55, 65, 66, 68, 70, 71, 88, 134, 136, 158, 176, 207, 208
learning skills, 68, 145
learning styles, 3, 9, 24, 30, 126, 128, 134, 205
learning task, 6, 80, 119, 127, 128, 131, 132, 159
legal issues, 199
legal systems, 199
lens, 22
lifelong learning, 108
Likert scale, 138
limitations, 37, 38, 110, 118, 140, 145, 177, 199
Lincoln, 99, 113
line graph, 25, 32
linear, 47, 123, 124, 125, 126, 138
linear model, 138
links, 48, 50, 114, 137, 196, 202
liquids, 77
listening, 120
literacy, 46, 64, 89, 112, 191, 193, 198
live surgery, 175
loading, 104
location, 170, 176
logging, 41
London, 33, 60, 61, 111, 112, 115, 165
long-distance, 168, 176
longitudinal study, 97, 98
long-term memory, 2, 3, 6, 12, 118, 119, 121, 123
low-income, 107
lumen, 176

M

machines, 38, 41, 42, 43, 121
mainstream, 193
maintenance, 5, 111, 144
management, 10, 34, 88, 139, 151, 159, 201, 210
Mandarin, 97
manipulation, 5, 123
MANOVA, 103, 105, 106
manufacturing, 180
market, 180
Mars, 81, 82, 122, 131
Maryland, 91
Massachusetts, 60
material resources, 1
mathematical skills, 18, 20
mathematics, 41, 97, 112, 114, 123, 126, 128
mathematics education, 41
measurement, 140, 141, 146
measures, 96, 99, 103, 106, 120, 123, 126, 133, 140, 141, 142, 144, 152
media, 8, 64, 129, 131, 134, 139, 145, 167, 196, 198, 200, 208
median, 98
mediation, 184, 185
memory, 2, 3, 6, 7, 8, 11, 12, 17, 27, 29, 75, 118, 119, 120, 121, 123, 124, 125, 129, 137
men, 154
mental activity, 67, 71
mental model, 59, 123, 132
mental processes, 118
mental representation, 8, 124
mental simulation, 123
mentor, 175
mentoring, 174, 176
messages, 19, 78, 79, 155, 184
meta-analysis, 131, 134, 144, 146
metacognition, 163
methodological procedures, 21, 29
metropolitan area, 110
Microsoft, 23, 24
middle class, 81
middle schools, 93, 136
military, 180, 203
Ministry of Education, 138
Minnesota, 59, 98
minority, 49, 107
miscommunication, 184
misconceptions, 3, 10, 11, 22, 27, 29, 31, 59, 60, 106, 200

missions, 82, 202
misunderstanding, 29
MIT, 60
mobile device, 156, 157, 160, 171, 175
mobile phone, 153, 158, 160, 180, 188
modalities, 118, 119, 120, 168
modality, 8, 35, 36, 120, 121
modeling, 91, 128
models, 36, 37, 44, 55, 59, 61, 77, 95, 109, 123, 127, 128, 132, 170, 173, 174, 175, 176, 202
modules, 46, 202, 209
mood, 154, 155
motion, 5, 35, 46, 47, 50, 53, 54, 122, 123, 126, 205
motivation, 1, 2, 12, 14, 15, 16, 17, 30, 56, 111, 184, 204
motor skills, 145
movement, 5, 135, 137, 193
multicultural, 9
multidimensional, 101
multidisciplinary, 100
multimedia, 1, 4, 6, 7, 8, 9, 19, 20, 30, 31, 33, 34, 35, 60, 117, 118, 120, 121, 127, 128, 130, 131, 133, 134, 135, 137, 138, 139, 140, 143, 144, 145, 146, 147, 167, 168, 170, 171, 177, 185, 186
multimedia data, 171
multiple-choice questions, 139
multivariate, 106, 114
multivariate statistics, 114

N

Nanyang Technological University, 201
National Research Council, 3, 34
natural, 6, 65, 94, 167, 168, 198
negative consequences, 117
negative influences, 181
Netherlands, 112, 115, 162
network, 40, 64, 108, 109, 110, 124, 125, 136, 154, 159, 162, 170, 172, 180, 182, 184, 185, 203
networked computers, 153
networking, 38, 162
New Jersey, 60, 146
New Orleans, 113, 114
New South Wales, 117
New York, 32, 34, 35, 59, 91, 111, 112, 114, 115, 128, 129, 130, 131, 146, 162, 165, 188, 189
Newton, 47
next generation, 210
niche market, 180
normal, 38, 47, 49, 53
normalization, 104
North America, 183, 192
North Carolina, 59
novelty, 38
nurse, 175
nutrition, 144

O

obese patients, 176
obesity, 135
observations, 3, 31, 41, 56, 65, 93, 98, 99, 100, 172, 181, 182, 183, 197
Oceania, 171
Ohio, 98
online communication, 153
online learning, 154, 155, 163, 208
operator, 172
optics, 172
organic, 129
organizational guidelines, 184
orientation, 47, 86, 97, 136, 174, 175
oscillations, 46, 47, 51
overload, 40, 118, 119, 120, 125

P

packets, 162
paradox, 181, 182
paradoxical, 181, 183, 184, 187
Paris, 65, 71, 88, 91
partnership, 171
passive, 2, 3, 152, 192, 198
patient care, 176, 177
patients, 168, 171, 173, 174, 176
PCs, 22
PDAs, 157, 171
pedagogical, 57, 63, 65, 71, 78, 79, 92, 94, 95, 100, 101, 108, 110, 201, 202, 203, 205, 207, 209, 210
pedagogy, 31, 39, 40, 43, 57, 94, 99, 100, 107, 114, 186, 187, 203, 204
peer, 29, 58, 109, 110, 155, 168, 185, 191, 195, 197
peer review, 168, 197
peers, 19, 26, 27, 28, 29, 30, 65, 108, 152, 156, 193, 194, 207
pendulum, 47, 50, 51, 52, 53, 54
perceived control, 129
perception, 5, 204

perceptions, 33, 94, 95, 98, 99, 101, 103, 106, 111, 113, 173, 201
peritoneum, 176
personal learning, 162
personal relevance, 107
personal responsibility, 30
Perth, 93, 112, 113
pH, 12, 19, 35
philosophy, 68
phone, 160
photographs, 40, 210
physical activity, 135
physical education, 134, 135, 139, 143, 145, 146, 147
physics, 25, 32, 33, 38, 46, 48, 49, 50, 53, 54, 55, 56, 57, 58, 59, 60, 146
pilot study, 138, 139, 140, 141
planning, 57, 76, 78, 79, 80, 86, 87, 96, 124, 168, 173
plastic, 109
platforms, 10, 153, 154
play, 31, 63, 64, 102, 134, 149, 152, 162, 177
PLC, 59
point-to-point, 171
police, 188
poor, 3, 39, 119, 171, 185, 192
population, 3, 85, 94, 108, 138
pornography, 182
portfolio, 156
Portugal, 163
positive correlation, 16
post-compulsory education, 49
power, 23, 43, 68
prediction, 120, 174
predictor variables, 96
preference, 8, 17
preservice teachers, 99
press, 34, 53, 97, 114, 178, 189
pressure, 21, 38, 122
prevention, 60
printing, 113
prior knowledge, 7, 8, 12, 21, 31, 42, 60, 80, 117, 121, 122, 123, 124, 125, 128, 129, 130, 131, 132, 204
private schools, 110, 180
probability, 138
problem solving, 34, 64, 87, 89, 90, 95, 121, 126, 152, 183, 184, 186
problem-based learning, 91, 132
problem-solving, 65, 68, 88, 91, 114, 119, 126
problem-solving skills, 91
procedural knowledge, 5
procrastination, 188
production, 46, 128, 177, 198, 199
productivity, 111, 112
professional development, 1, 93, 94, 97, 106, 113, 187
professional qualifications, 46
program, 5, 22, 24, 25, 56, 93, 94, 96, 97, 99, 100, 101, 103, 106, 107, 108, 109, 110, 113, 114, 127, 137, 138, 139, 144, 174, 175, 176, 186, 202, 205, 207, 208, 209
programming, 128, 158
projector, 40
proposition, 37
protocol, 50, 52, 184
psychologist, 45, 204
psychology, 31, 80
public, 46, 95, 100, 102, 106, 107, 109, 110, 163, 181, 198
public sector, 46
pupils, 39, 40, 41, 42, 43, 44, 55, 57, 58, 61, 90, 107, 114, 152

Q

qualitative differences, 124
quality control, 39
quanta, 45, 60
quantitative research, 129
quantum, 46, 56
quantum phenomena, 46, 56
Quebec, 163
query, 90
questioning, 100, 115
questionnaire, 101, 138, 141, 143
questionnaires, 95, 209
quizzes, 32, 157

R

radiation, 43, 47, 56
radiation detectors, 47
radio, 205
random, 97, 118, 136
range, 23, 39, 41, 42, 43, 46, 47, 49, 72, 88, 109, 170, 171, 183, 185, 194, 198
rating scale, 120, 121
ratings, 122, 126, 127, 210

reading, 11, 42, 48, 52, 75, 80, 91, 92, 121, 129, 146, 153, 181
reading comprehension, 91, 129
real time, 41, 126, 154, 162, 173
realism, 122, 174, 177
reality, 4, 153, 155, 167, 168, 176, 177, 180
reasoning, 48
recall, 130
recalling, 12
reciprocity, 183, 185, 186
recognition, 2
reconstruction, 167, 173
recreational, 174
Redox, 19
redundancy, 119, 120
reflection, 36, 52, 79, 100, 108, 154, 159, 180, 195
reforms, 94
regeneration, 59
regional, 98, 173
regression, 112
regular, 54, 144, 145, 171
regulation, 68, 129, 196
regulations, 74, 75
Reimann, 35
reinforcement, 139
relationship, 53, 107, 131, 173, 174
relationships, 9, 75, 96, 97, 159, 184, 204
relaxation, 146
relevance, 28, 29, 47, 94, 96, 101, 107, 173
reliability, 97, 98, 104, 105, 112, 138, 140, 141, 174
Research and Development, 35, 90, 129, 130, 131, 132
research design, 94
resection, 175
resistance, 22, 23, 52
resistivity, 47, 56
resolution, 78, 88
retention, 32, 104, 133, 135, 140, 144, 145
Reynolds, 135, 144, 146
risk, 56, 102, 162
robotics, 173
Robotics, 167
robustness, 7
role-playing, 34
routines, 156
Royal Society, 35, 60

S

safety, 2, 74, 75, 176
sample, 26, 27, 85, 97, 98, 99, 104, 110, 144, 174
sampling, 136
satisfaction, 33, 144, 154
scaffold, 45, 71, 73, 78, 79, 93, 100
scaffolding, 38, 44, 45, 48, 64, 66, 67, 68, 70, 71, 80, 88, 89, 90, 124, 132, 181
scaffolds, 63, 64, 65, 66, 70, 71, 72, 73, 74, 75, 76, 77, 78, 79, 92, 124
scheduling, 167
schema, 7, 8, 11, 12, 17, 30
schemas, 118
scholarship, 192, 193, 196
school climate, 112
school learning, 93, 95
school work, 53
schooling, 110
science education, 1, 2, 3, 5, 31, 40, 41, 56, 59, 61, 63, 66, 94, 106, 107, 108, 109, 112, 115
science educators, 43, 95, 109
science teaching, 37, 38, 43, 44, 56, 63, 64, 110, 112, 115
scientific knowledge, 64, 65, 67, 87, 96, 168
scientific method, 55, 108
scores, 84, 85, 87, 97, 98, 103, 123, 127, 135, 138, 139, 140, 142, 143, 144
search, 39, 59, 72, 73, 74, 77, 80, 82, 83, 85, 86, 87, 88, 90, 92, 118, 120, 124, 160, 186, 207
search engine, 72, 73, 74, 80, 186
searches, 73, 86
searching, 63, 66, 73, 76, 79, 80, 82, 83, 86, 87, 88, 89, 90, 91, 124
second language, 119
secondary education, 63, 64, 66, 80, 81, 82, 86, 89
secondary school students, 77
secondary schools, 60
secondary students, 80, 86, 88, 91, 101
selecting, 39, 51, 65, 67, 75, 78, 80, 86, 87, 95, 101
selective attention, 132
self-assessment, 19, 20
self-esteem, 71
self-paced learning, 171
self-reflection, 184
self-regulation, 68, 129
self-report, 98
semantic, 125, 159
sensors, 6, 24, 41
sentence comprehension, 126
sequencing, 71, 78, 119
series, 24, 27, 49, 68, 69, 70, 71, 72, 74, 75, 78, 119, 121, 126, 180

services, 152, 154, 158, 159, 161, 170, 203
sesame, 188
shame, 41
shape, 44, 102, 170
sharing, 30, 156, 168, 171, 184, 198
short period, 2
short-term, 17, 44, 129
shoulder, 147, 148
signals, 21
simulation, 1, 3, 4, 5, 11, 12, 13, 15, 16, 17, 18, 20, 21, 22, 23, 24, 25, 26, 30, 32, 33, 34, 35, 49, 76, 122, 123, 137, 167, 170, 173, 174, 177
Singapore, 96, 111, 112, 113, 201, 211
sites, 39, 72, 161, 171, 184, 192
SMEs, 202, 205
SMS, 156
social construct, 9, 152
social constructivism, 152
social context, 61
social network, 153, 180, 184, 185, 207
social presence, 5, 33, 154, 161
social problems, 199
social psychology, 125
social sciences, 114
software, 34, 38, 43, 44, 48, 50, 53, 54, 56, 58, 134, 135, 145, 147, 156, 160, 164, 173, 180, 185, 207, 208
sounds, 196
South Africa, 97, 111
Spain, 63, 81, 165
spam, 184
specialization, 94
specific knowledge, 123, 124, 132
specificity, 124, 145
spectroscopy, 19, 25, 34
spectrum, 26, 27, 167
speech, 79
speed, 38, 41, 42, 172
sports, 135, 145, 182
spreadsheets, 12, 24, 25, 75
SPSS, 111, 146
SQL, 202, 208
stages, 41, 47, 48, 69, 72, 78, 119, 126
standard deviation, 84, 97, 103, 105, 106, 140, 141
standardization, 168, 174
standards, 113, 153, 172, 192, 194, 198
Standards, 34, 60, 102, 113
statistical analysis, 12, 17, 19
statistics, 66, 90, 103, 111, 114
stimulus, 21
storage, 6, 38, 129
strategies, 30, 44, 59, 63, 64, 65, 66, 68, 69, 75, 79, 80, 82, 83, 86, 87, 88, 89, 90, 91, 92, 95, 98, 108, 114, 115, 126, 145, 146, 173, 183, 184, 185, 186, 187, 203
strategy use, 78, 80
streams, 168
strength, 28, 47, 52, 56
stress, 20
structuring, 45, 68, 125, 132
student achievement, 32, 64, 124, 132
student feedback, 185, 187, 195
student motivation, 184
subjective, 120, 122, 126
substitutes, 148
subtasks, 83
suburban, 108
summer, 94, 99, 100, 103, 109
superiority, 135, 174
supervision, 21, 43, 44, 47, 57, 58, 139, 204
supervisor, 111
supplemental, 145
supply, 108, 173, 175
surgeons, 168, 171, 172, 173, 174, 175, 176, 177, 178
surgeries, 168
surgery, 167, 168, 170, 172, 174, 176, 177, 178
surgical, 167, 168, 170, 172, 173, 174, 175, 176, 177, 178
surgical intervention, 171
synchronous, 4, 10, 29, 36, 203, 207
syndrome, 156, 188
Synergy, 33
synthesis, 112, 184

T

Taiwan, 99, 111
tangible, 94, 193
tanks, 107
targets, 162
task difficulty, 126
taste, 134
teacher preparation, 94
teacher support, 56
teaching abilities, 161
teaching experience, 184
teaching process, 135, 146
teaching scenarios, 162
teaching strategies, 95, 145, 186, 187, 203

teaching/learning process, 70, 71
technological revolution, 168
teenagers, 80, 192
telecommunications, 167, 171
telemedicine, 171, 177
television, 50, 51, 54
temperature, 21, 23, 42, 82, 83, 122
temporal, 39, 120
tension, 175
test items, 138
test procedure, 139
test scores, 84, 85, 133, 136, 138, 141, 142, 143, 145
test statistic, 12, 14, 15
Texas, 93, 94, 96, 98, 108, 113
text messaging, 188
textbooks, 40, 44, 191, 193, 196, 205
theft, 182
Thessaloniki, 136, 144
thinking, 2, 3, 29, 31, 36, 43, 44, 48, 49, 50, 51, 52, 53, 54, 57, 58, 65, 70, 78, 80, 195
time frame, 160
time periods, 54
time use, 145
titration, 11, 12, 17, 19
toxic, 74
trade, 46
tradition, 95, 155
traffic, 127, 129, 131, 192
trainees, 120, 172, 174, 175, 176, 201, 202, 204, 205, 207, 209, 210
training, 12, 18, 19, 20, 28, 31, 33, 46, 87, 90, 91, 94, 108, 111, 120, 121, 126, 127, 129, 130, 131, 135, 136, 143, 167, 168, 170, 172, 173, 174, 175, 176, 177, 178, 188, 201, 202, 207, 209
transfer, 34, 36, 40, 64, 91, 94, 95, 99, 111, 127, 131, 160, 168
transference, 9
transformation, 5, 99, 181, 199, 201
transition, 5, 35
transitions, 60
translation, 5, 170
transmission, 9, 172, 173, 174, 176, 177, 182
transparency, 173
transport, 74, 203
transportation, 181
travel, 175
trial, 48, 177
trustworthiness, 198
tuition, 180, 182
tutoring, 92, 128
two-dimensional (2D), 173
two-way, 172, 203
typology, 60, 79

U

undergraduate, 32, 179, 180, 182, 183, 185, 186, 188, 192
undergraduate chemistry course, 32
undergraduate education, 179
uniform, 47, 117, 153
United Kingdom, 37
United States, 102
univariate, 106
universal law, 47
universities, 10, 110, 151, 180, 181, 182, 183, 192
university education, 187
university students, 31, 40
updating, 155, 168
upload, 194, 198, 205, 208, 210

V

validation, 111, 145, 174, 177
validity, 56, 97, 98, 101, 104, 112, 138, 174
values, 5, 19, 23, 26, 43, 83, 102, 104, 122, 140, 184, 187, 203, 208
variability, 124
variables, 5, 10, 16, 17, 30, 66, 74, 75, 78, 79, 84, 85, 96, 109, 140
variance, 104, 105, 140
vascular surgery, 171
vector, 126, 137
Victoria, 113
videoconferencing, 167, 171, 172
Videoconferencing, 171, 172
village, 109
virtual instruments, 12
virtual reality, 167, 168, 177
virtual university, 168
viscosity, 34
visible, 40, 204
vision, 28, 108, 201
visual perception, 210
visualization, 5, 34, 173
vocabulary, 137
volleyball, 134, 135, 136, 137, 138, 139, 143, 144, 145, 146, 147, 148
Vygotsky, 45, 152, 165

W

Wales, 117
warfare, 181
water, 47, 56, 199
weapons, 181
weapons of mass destruction, 181
wear, 185
web browser, 82
web pages, 74, 75
web sites, 72
web-based, 1, 31, 63, 64, 66, 67, 68, 69, 70, 71, 72, 73, 80, 81, 82, 83, 86, 88, 128, 188, 191, 202, 208
web-based instruction, 66, 81, 83
Weblog, 156, 159, 163
websites, 108, 168, 171, 192, 194, 196
Wi-Fi, 156
windows, 208
wireless, 162, 176, 178, 180, 181, 188
withdrawal, 50, 79
word processing, 146
working hours, 168
working memory, 2, 3, 6, 7, 8, 12, 29, 118, 119, 120, 124, 125
workload, 181, 197
World Wide Web, 25, 64, 65, 80, 88, 89, 90, 151, 152, 154, 158, 160, 180
writing, 77, 78, 92, 146, 159, 163, 185, 191, 192, 193, 194, 195, 196, 197, 199, 205
writing process, 197
writing tasks, 199

Y

yield, 145
young adults, 188